Pilot in Command

Other McGraw-Hill Books of Interest

Pilot in Command

Paul A. Craig

McGraw-Hill

New York   San Francisco   Washington, D.C.   Auckland   Bogotá
Caracas   Lisbon   London   Madrid   Mexico City   Milan
Montreal   New Delhi   San Juan   Singapore
Sydney   Tokyo   Toronto

Cataloging-in-Publication Data is on file with the Library of Congress

McGraw-Hill

A Division of The McGraw-Hill Companies

Copyright © 2000 by The McGraw-Hill Companies, Inc. All rights reserved. Printed in the United States of America. Except as permitted under the United States Copyright Act of 1976, no part of this publication may be reproduced or distributed in any form or by any means, or stored in a data base or retrieval system, without the prior written permission of the publisher.

1 2 3 4 5 6 7 8 9 0   DOC/DOC   9 0 4 3 2 1 0 9

ISBN 0-07-134844-1

The sponsoring editor for this book was Shelley Carr, the editing supervisor was Steven Melvin, and the production supervisor was Pamela Pelton. It was set in Garamond per the Gen1AV1 design by Joanne Morbit and Michele Pridmore of McGraw-Hill's desktop publishing department, Hightstown, N.J.

Printed and bound by R. R. Donnelley and Sons, Inc.

McGraw-Hill books are available at special quantity discounts to use as premiums and sales promotions, or for use in corporate training programs. For more information, please write to the Director of Special Sales, McGraw-Hill, 11 West 19th Street, New York, NY 10011. Or contact your local bookstore.

 This book is printed on recycled, acid-free paper containing a minimum of 50% recycled, de-inked fiber.

In the winter of 1998 I went to the NASA Ames Research Center in California to participate in a planning session. The NASA Aviation Safety Program had been given the charge to: "Improve aviation safety five-fold over the next 10 years and a ten-fold improvement over the next 20 years." Even though the aviation accident rates are very low compared to other forms of transportation, it is feared that even if accident percentages remain the same while total flying increases, the total number of accidents will go up. General aviation has the most to lose because it is the largest segment of aviation and at the same time has the higher accident rates. I hope that this work will be of some help toward the NASA and national goal of improving safety by reducing accidents and saving lives. Specifically, this book is dedicated to the general aviation pilots who may more clearly understand their role as pilot in command and therefore fly safer, fly with more confidence, and fly with more fun.

Franklin, Tennessee
June 1999

Contents

Introduction

On June 1, 1999, an American Airlines MD-80 crashed off the far end of a runway in Little Rock, Arkansas. The National Transportation Safety Board (NTSB) and the media began to investigate the accident. There were questions about the airplane's control surfaces and thrust reversers, but the central question became "Why did the pilot land at Little Rock in the face of what was later determined to be a level 6 thunderstorm?" Recordings and interviews revealed that the pilots were aware of the storm but elected to land anyway. Two days after the crash, and after some of the facts were coming out, I was watching the reports of the accident on CNN. I heard the reporter say, "It's hard to believe, but a controller cannot tell a pilot to land or not to land their airplane, even when the controller knows about a storm." This statement, broadcast to the world, is the result of an almost universal misconception. Controllers do not control airplanes, pilots do. It is one thing to have people in the media misunderstand, but the bigger problem is that many pilots also do not understand. Many pilots fly as pilot in command but are not completely aware of the tremendous authority and responsibility that come with being the pilot in command. A 16-year-old student pilot flying for the first time solo has the decision authority over the most senior and experienced air traffic controller in the country. The misconception of this fact has had a direct influence on safety and accident rates. Pilots in general aviation especially need to reposition their view and understanding of what it means to be pilot in command.

This book first redefines pilot in command as something more than what is written in a logbook. If pilots, in fact, have an unrealistic view of being pilot in command, how can we start teaching what it is really like? The book points out that there is a difference between general aviation and airline training methods and asks the question "What are the airlines doing that we in general aviation should be doing?"

It is true that the greater experience and expertise pilots have, the less likely it is that they will be involved in an accident. So one way to have fewer accidents is to have more pilots perform like experts. How can nonexpert pilots act like experts? They can start doing what experts do. This book details what expert pilot behavior is, and then demonstrates how nonexperts can move toward expert performance. The book then outlines and gives examples of how these techniques can be put into practice during everyday flight training. Finally, the book's Appendix describes a year-long research project involving general aviation pilots who were exposed to "airline-type" decision training. This research was the first of its kind aimed at general aviation pilots in over 25 years.

Acknowledgments

No book is the work of a single person and I have many debts and appreciations for this one. I spent over 3 years in the preparation, research, and writing of this book, and my wife has put up with it all. Special thanks and love then go to Dr. Dorothy Valcarcel Craig.

Drs. Charles Dickens, David McCargar, R.O. Renfro, and John Bertrand of Tennessee State University deserve many thanks for ideas, mentoring, and encouragement.

My assistant for the flight simulator sessions was Lauren Bandy. Lauren graduated with honors from the Department of Aerospace at Middle Tennessee State University while working on this project. He is now an airline pilot, but he remains a skilled and caring instructor and researcher.

Great friends and aviation professionals Captain Ken Futrell of United Airlines, Joel Smith of Northwest Airlines, and Phil Heitman, of Dreamflyer Publications have always been a source of help and laughs.

Larry Lambert, the North Carolina Safety Program Manager; Roger Anderson, the Nashville Air Traffic Support Manager; and Robert Cope of the Nashville FSDO have offered information and support on many projects, this one included.

I continue to be surrounded by a team of professional flight instructors who daily honor their commitment and understand what a privilege it is to be a flight instructor: Tracy Whitt, Chris Allen, Servando Gomez, Shawn Collins, Daniel Lundberg, Ashley Neal, Josh Hamilton, Mike Robison, Chris Hollins, Kim Osby, Erik Holmbjörk, Lance South, Eric Lorvig, Jenny Troutner, Jeff Johnson, Jonathan Frase, Jonathan Reeves, Matt Striegel, Brannon Daugherty, Roger Justice, Antonia Hayes, and Matt Terrier.

A extra special thanks to Joyce Maynard, my working partner; Terry Dorris, my Cessna 404 and simulator partner; and all the faculty, staff, and students at Middle Tennessee State University Aerospace.

Pilot in Command

1

The Responsibility of Pilot in Command

The rule that defines *pilot in command,* may be the most straight-forward regulation in the entire FAA regulation book. FAR 91.3(a) says *The pilot in command of an aircraft is directly responsible for, and is the final authority as to, the operation of that aircraft.* Other rules are longer, but when all the legal jargon is cut away, they often say very little. This regulation, however, is short on words but large in meaning.

I read this regulation first as a student pilot and at various other times over the years, but now I realize that I was actually reading past it. I did not grasp the true implications of that one sentence. I thought that being "in command" meant that it was up to me as the pilot to decide when to turn base in the traffic pattern or how fast to taxi or what altitude to climb to. Of course, the pilot in command is responsible for those decisions but is also responsible for much more. The pilot in command is a planner. The pilot in command creates the flight's destiny. Yes, it often seems that we are not in control. It seems that we are at the mercy of the regulations, the cost of flying, and the air traffic controllers. Granted there are times when the regulations and the cost will dictate our actions. Granted there will be times when the air traffic controller does seem more in control, but there is no such thing as "controller in command." Ultimately the pilot must evaluate the controller's suggestions and either accept or reject the instructions. Pilots must use good judgment here. If a controller gives the instruction to make an immediate turn to avoid traffic, it is the controller who is in the best position to make a call, and I highly advise taking that suggestion. However, someday you may be assigned an altitude that places you inside an icing layer. Usually icing layers are not very thick, and a climb or a descent could solve the problem. First mention to the controller that you are picking up ice and need another altitude. If an altitude change is not forthcoming and the ice becomes greater than you are comfortable with, announce to the controller what altitude you are changing to. You are in command.

You are also in command when it comes to routing. I have received route changes that would take the flight through severe weather. But there is not a controller alive who can get me to fly through a thunderstorm. I will ask for deviations, and I will tell the controller when and if it will be possible to rejoin the original course. And, of course, you are in command when it comes to unplanned or precautionary landings. When there are IFR conditions, ice, turbulence, or headwinds, or simply a rest room stop is needed, you are in control. A controller cannot suggest to a pilot where to deviate in the event of an unplanned landing. You announce your intentions to the controllers, and you fly your airplane where you want to go. But unfortunately, many pilots who routinely act as pilot in command are not comfortable taking command. Many pilots are timid when assertive action is needed. They do not want to get "in trouble," so they do nothing.

One of the best learning experiences you can ever have is to take a trip to your local control tower and radar room. Call ahead and for small groups and with a little warning they will be happy to show you around. The first thing you will notice on your visit is that these people are just regular folks. They usually are dressed casually, and the atmosphere is surprisingly relaxed. You will quickly see that controllers are professionals who have one goal: to get airplanes on the ground and go home. They do not want to pick a fight with pilots. They are not sitting there like the highway patrol with a radar gun and a citation book. The next time you fly after your visit you will have a clearer understanding, and you will not be as timid. After all, the pilot's job is also to get the airplane on the ground and go home. You will feel more at ease speaking on the radio and directing the flight's destiny after a trip to the tower. After all you are just carrying on a conversation with people who have the same goals as you do.

In fact, controllers would much rather the pilot be the leader. For liability reasons there are many items that a controller cannot suggest or recommend to a pilot. The controller is taken off the hook when the pilot really is in command. I have spoken to and worked with many controllers. They can recount situations where pilots were not assertive and depended on the controllers to recommend their course of action. Sometimes pilots require such hand holding that when it's all over, the controllers often feel that they deserve to log the flight as pilot in command. The pilot is supposed to be in control, but when the pilot does not take control, this places the controller in a very awkward situation.

I was flying an ILS approach through the clouds one time and after being switched over to the control tower frequency, I heard another pilot say, "Tower this is N1234A, 5 miles south, inbound for landing." The controller responded, "The field is IFR at this time, what are your intentions?" Now this conversation really got my attention because I was still descending in the clouds and for all I knew this VFR pilot was just under me at the base of the clouds. The "What are your intentions?" question confused the VFR pilot. He first did not understand that he was being denied entry to the airport. The pilot repeated the request, this time with a clear urgency in his voice, "I need to land." The solution to the problem was for the pilot to ask for a special VFR clearance, but the pilot did not ask, and it was clear to all listening that he probably did not know that he should. On the other hand the controller could not assign the special VFR clearance to the pilot. What if the controller had told the pilot to fly to the airport using this "special" clearance and while en route the pilot flew into a cloud, lost control, and crashed. The controller would be accused by the lawyers later of putting the inexperienced pilot into a situation that the pilot could not handle. So the controller could not tell and the pilot did not know to ask. This is the awkward and dangerous situation that can develop when a person flying around in an airplane is not ready to serve as pilot in command. Then the controller did a brilliant thing. He said to the nervous pilot "a few minutes ago another VFR pilot got in when he asked for a *special VFR*. What are your intentions?" The pilot then did say the words *special* and *VFR* in the same sentence, although I was never convinced he really knew what was going on. The special VFR was granted, the pilot landed safely, and I landed on a different runway.

Once an air traffic controller was sued by the surviving relatives of a pilot. The pilot died in an accident in which he flew his airplane into a thunderstorm. The pilot had been in communication with a center controller about the weather ahead. The pilot was seeking a path through a line of precipitation, and the controller had suggested a heading that appeared to aim the airplane at the narrowest band. The civil action was taken against the controller for suggesting the heading. The family contended that by telling the pilot to fly the heading the controller was placing the pilot into danger. They asserted that it was the controller's decision that led to the death of the pilot. But the court held that the pilot was the one in command and that the suggested heading given by the controller was just that—a suggestion that could have been either accepted or rejected. In essence the court said that there is no such thing as controller in command and that *pilot in command* means what it says.

Of course, many flights take place without interaction with air traffic controllers, and being in command is necessary on every flight. Many pilots associate this complete state of command with the freedom of flight. When else in our lives can we say that we are completely in command of anything? I am never completely in command except in an airplane. On the ground it seems everyone has hooks in me. I am certainly not totally in charge at work. My employer and colleagues all have demands on my time. I have deadlines to meet, reports to write, classes to teach, and students to advise; I am always on the run. I have bills to pay, a dentist appointment, committee meetings, civic club luncheons, grass to mow, a car that needs an oil change, kids to pick up. I can't wait to get in the airplane and shut the door. That is the only place where I can really get organized and where what I say goes. What a great oasis we pilots have. When you do accept the responsibility of being pilot in command, it can be very empowering.

But pilots must compartmentalize. We cannot take all those demands and problems into the air with us. We must focus once we approach the flight so that we are not thinking about paying the bills when we should be thinking about the crosswind. A nonpilot friend of mine gave me a great compliment once without knowing it. He said that I was always joking and fooling around except when I got around an airplane that I intended to fly. Then I was all business. One of the reasons pilots love to fly is because flying can be a great escape. Flight is freedom from all those concerns on the ground. I love being pilot in command.

Back to the regulation. This law makes an assumption that pilots flying aircraft know what they have gotten themselves into and that they in fact know how to be in command. I flew for many years without actually knowing this. I read these words, but I guess I really did not believe them. I did not really feel empowered by them. I felt subservient to FSS weather briefers, air traffic controllers, examiners, and other pilots. I was passive. I was not assertive. I was flying, but I was not actually in command. I flew in a shell, too afraid to take control.

I would have never asked a controller a question on the radio for fear that the question would be out of place or stupid. My lack of experience and low pilot confidence prevented me from speaking up around pilot examiners, not to mention the all-mighty Federal Aviation Authority (FAA). It was a cycle that was hard to break. I knew I was not a complete pilot, yet I could not get over the hurdle for fear of falling on my face. Flying without confidence is uncomfortable, so I began to shy away from flight situations that required confidence. I hardly ever flew anywhere that required radio communications. When I did, I was tentative and felt that I was intruding in the world of the pros. I was in the way, in over my head, and miserable. On top of all that, I was simply not safe. Like most pilots, I did not experience any real emergencies. The flights seemed to always work out, but I knew that I was biding my time. Someday something would happen, and I would be completely unprepared to face it. I would not have been able to be in command in a crunch. Flying really was not much fun for me unless the stakes were very low. Otherwise I had the feeling that I was walking on thin ice. I was more afraid of falling through than eager to get to the other side. When I read the pilot in command regulation, it did not seem to have power for me. After all, how could I be the "final authority" when I did not feel in control to begin with? Something had to change.

Something did change. Years passed and I gained experience. I scared myself many times, but somehow I learned from mistakes without killing myself. Others have not been so lucky.

There is a danger zone that all pilots must traverse. It starts on the day they get their first pilot certificate and can go out on their own as pilot in command and proceeds to a day when experience becomes their guide. This trip across the danger zone does not end on any specific day or with any specific number of flight hours in a logbook or any specific number of years since the checkride. No, this journey has no specific stopping point and takes place gradually. Some pilots never make it all the way across. Some stop flying, some never learn, and some get killed in the process. And this danger zone is not some theoretical concept. It is actually statistical and can be clearly seen on a chart.

A view of accident statistics proves that flying is safe but not without risk. When comparing general aviation with other categories of flight operations, a gap can be seen. Using accidents for every 100,000 hours as the basis for comparison, it can be seen that air taxis, commuter airlines, and major air carriers have far fewer accidents than pilots in general aviation (Fig. 1-1). True, general aviation represents a wider variety of aviation activity. Banner towing, power line patrol, air ambulance, and even flight instruction can offer more risk than simply flying from one point to another, but it is also clear that general aviation pilots on average have less experience than airline or military pilots.

It is a fact that pilots with more experience have fewer accidents. It is a classic case of the rich getting richer while the poor get poorer. Pilots who lack experience have more accidents while making the attempt to gain experience. So the challenge is clear: Is it possible to partly substitute education for experience, at least in the short run? What can be done to help pilots who are crossing the danger zone to have fewer accidents? In other words, are there teaching and learning strategies that can be used that will "train out" the next accident? This would keep inexperienced pilots safer and would afford them the opportunity to gain experience in the long run.

This is not a how-to book on things for low-time pilots to do to remain safe. That would be all together too simplistic. I believe that

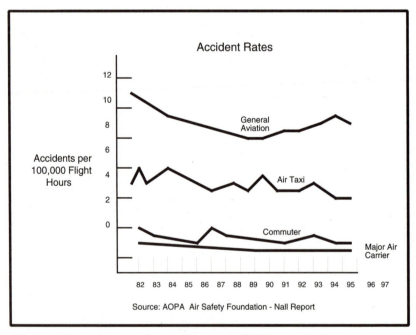

Fig. 1-1 *Comparison of accident rates of different flight-operation categories.*

the problems are vastly more complex. It would not be possible to dream up all scenarios that a pilot might face and offer a suggested course of action (although some examples are given in later chapters). Instead, this book is about how to better prepare to make the decisions as they are faced. It also looks at the results of our aviation training and how systems have developed that actually can "train in" accidents. Are we training pilots to simply perform a series of maneuvers, or are we teaching decision makers to function as members of a critical system?

In the past, the explanations given for the causes of accidents have been way too simple to meet the reality test. The phrases "get-home-itis" and "get-there-itis" have been used to explain an accident scenario. The story is familiar to you. A pilot is anxious to get home from or get started on a weekend trip. The pressure to get home or to stay on a schedule clouds a good pilot's judgment. The pilot, who knows better, flies into conditions that he or she is not prepared for and an accident takes place. The reason for the accident is neat and tidy, and all who hear it are sad for the loss but happy that the accident could be so easily explained. It was "pilot error." Actually, accidents are just not that simple. We should be asking the tougher, deeper questions. Why did the pilot believe that his or her actions were safe when in fact they were not safe? What examples had been set for the pilot? Did the pilot ask for help? Did the pilot know how to ask for help? Did the pilot even know that help was needed? Was the pilot afraid to ask for help? Was the pilot afraid to declare an emergency because he or she did not want to get into trouble with the FAA? Had the pilot been taught to deal with outside pressures? Had the pilot been taught to be pilot in command? Writing the accident off as a product of simple get-home-itis illustrates a lack of understanding. The problem did not take place in a vacuum involving one person at one place at one time. Instead the accident was the culmination of many complex and interrelated factors. Get-home-itis takes everyone except the pilot off the hook. Providing a simple answer to this complex problem has the effect of shifting the accident blame to only the pilot and ignores the training, the regulations, the attitudes, and the environment that the pilot grew up in.

Can a pilot be held solely responsible for an accident when it is the flight instructor who teaches habits and attitudes? Can the pilot be held solely responsible when the FAA sets the rules for obtaining a pilot certificate that places more emphasis on the performance of maneuvers than on decision making? Can the pilot alone be blamed for the problems of the entire system? Get-home-itis is a simple explanation for a complex problem and seems to provide cover for all involved except the person most hurt: the pilot.

Accident categories

Of course there are some easily explained accidents. I call them "pure" pilot errors. Some accidents do not have complex causes; they are the result of a single mistake made by the pilot. A pilot forgets to put the landing gear down and lands on the pavement instead of the wheels. A pilot fails to turn on the carburetor heat when flying in rain, and ice forms in the carburetor. A pilot loses track of position and inadvertently flies into Class B airspace. Pure pilot error is characterized by a single event that takes place at a critical time. The error happens because a pilot forgets to throw a switch or make a call or follow a procedure.

Another uncomplicated accident cause is mechanical. An accident can occur when something breaks. It is nobody's direct fault; it's just that a part or component simply stops working. The pilot tries to put the landing gear down, but the gear will not come down. A landing light burns out, and the pilot collapses the nose gear on touchdown because it was hard to see the ground. The vacuum pump fails, and the pilot becomes disoriented while trusting failed gyro instruments. The machine fails, and an accident is the result. Simple.

Simple pilot mistakes and simple mechanical failures do cause accidents, but not too many. The majority of accidents have more than one cause. In most cases the accident is the culminating event at the end of an event string that may reach back years into the past. The accident cause may have roots in the pilot's original training. Dozens of other factors could have also played a part. Combinations of mistakes, attitudes, habits, and misunderstandings construct the accident chain. These are "system" errors that accumulate over time.

The accident that happens is a symptom of the larger problem. Calling a culmination accident a pilot error is treating only the symptom and is short sighted. Culmination errors are bred within systems, so it is the systems and attitudes within the systems that need reforming.

A more realistic look

When airline pilots arrive at work to begin a 3-day trip, they very often meet their flight crew for the very first time. This always seems odd. They will spend the next 3 days making life and death decisions with total strangers. How is it possible for one pilot to depend on the other when they just met 5 minutes ago? Doesn't trust build over time? The answer is the use of standard operating procedure (SOP). If each crew member follows a standard of operation, each will know what to expect from the other even though the crew has no history with each other. When it comes to preflight, checklists, takeoff calculations, en route procedures, and approach and landing callouts, everything is by the book, and everyone has a specific job. When everyone takes care of his or her part of the job, the job always gets done. If anyone fails to follow the standard procedures, there will be a high risk of leaving something out, and the flight will be less safe. In general aviation we have SOPs as well, but ours are far less formal. We have training course outlines, checklists, and test standards, but often we allow familiarity and routine to erode our SOP.

I call this erosion *creeping normalcy,* and there are countless examples of this in our everyday lives. When I get in my car, I know that the power seat doesn't work anymore, and I must jiggle the key just so to make it start. When the engine does start, it will stall very often at red lights. Over time these problems stopped being nuisances and became just normal facts of life. When I drive the car, I think nothing of them, but I would have never bought the car in that condition. What I think is okay has changed. What is normal for me now was not normal once before. It was not a change that happened all at once, or I would have noticed it more. It was a gradual acceptance. What I considered to be normal slowly changed.

Checking the fuel sumps during a preflight inspection is another example. When you were first taught to test the fuel, you learned the importance of checking for uncontaminated fuel. You waited a sufficient length of time after fueling to allow any water that might be in the tanks to settle to the bottom. You drained a sample into the cup from the airplane's wing sump. Then you smelled the sample to ensure that it was in fact gasoline. Then you held the clear cup up against the white part of the airplane to see that the fuel's color tint was correct. You disposed of the sample. You did this 20 times before 20 different flights, 50 flights, 100 flights, and you never saw any water, and the fuel smell and color were always correct. One day you were in a hurry. You quickly sumped the fuel and discarded it without your usual care. The flight took place with no problems. The next flight, when you were not in a particular hurry, you again took the sample and discarded it. Soon your normal routine no longer involved smelling the fuel and checking its color. Without actually thinking about it, your standard had eroded. Not all at once, but slowly over time your SOP changed. What was not normal once became normal. Someday you will fill the fuel sample cup with condensation water and, thinking it must be fuel, you will discard it. Because you now have no SOP for determining if the sample is truly fuel, you will only discover that it is water after takeoff.

I completed a flight check with a student one time, and as we were leaving the airplane, I noticed that the pitot tube cover was still laying on the back seat. The student was on the side of the airplane that had the pitot tube, so I called to the student and told him that we had forgotten to install the cover. He said, "That's okay, the cover was not on when we first took the airplane." Do you see what had happened? Somehow over time it became normal for that student not to cover the pitot tube. There had been no apparent consequence for leaving it off, and it became routine not to install the cover. I wondered what else in this student's normal procedures had slowly changed. The problem seems to be that Murphy is not right. Murphy's Law says that whatever can go wrong probably will go wrong. But in actual fact most of the time what can go wrong will go right. Most of the time despite an inadequate fuel test, or a pitot tube left uncovered, the flight will take place with no problems. We get away with it. When we get away with it once and then twice, soon the relaxed standard becomes routine. What we consider normal shifts and usually there is no price to pay. Then one day the luck runs out, and the item that is no longer SOP causes an accident. Was that an isolated, one-time event, or was it the product of people and a system that allows creeping normalcy?

Several years ago a midair collision occurred that showed distinct signs of a shift in standard procedures on the part of both pilots and controllers. A Cessna 206 collided with an Air U.S. Jetstream over Loveland, Colorado. The Cessna was carrying skydivers up through 13,000 feet for a jump. The Jetstream, a commuter airliner, was en route from Denver to Gillette, Wyoming. When the two airplanes collided, two passengers in the Cessna were killed by the impact.

The pilot of the Cessna and three others were wearing parachutes at the time of the collision. Those four jumped free of the airplanes, deployed their parachutes, and survived the accident. All 13 people on the Jetstream were killed (Fig. 1-2). The visibility was reported to be 60 miles.

The pilot of the Cessna had flown many skydiver flights in that area and was climbing for his second jump of the day. At the time and at the location of the accident, a Mode-C altitude reporting transponder was required above 12,500 feet. The Cessna did not have a Mode-C transponder, but routinely conducted jump operations above 12,500 feet. In the past the controllers at Denver Center had granted a deviation to the Mode-C rule for the purpose of skydiving. Also the controllers on past jumps above 12,500 feet had frequently assigned the jump plane the transponder code of 1-2-3-4, and that was the code assigned on the first jump flight of the day. On the second flight the Cessna pilot again put the code 1-2-3-4 in the transponder and without communicating with Denver Center climbed through 12,500 feet. It was at 13,000 feet that the Jetstream overtook and collided with the Cessna. Several items had become normal to this pilot and controllers that should not have been. After using the same transponder code time and time again for skydiving, the pilot began to believe that this was a permanently assigned code and by merely using the code, it was a signal to the controllers that jump operations were under way. The controllers had routinely allowed the skydive

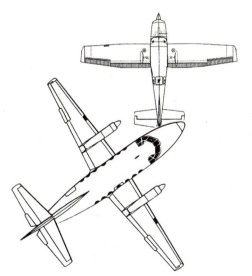

Fig. 1-2
Diagram of midair collision over Loveland, Colorado.

airplane to waive the Mode-C rule, so it became a normal situation to have the airplane operate without altitude reporting where it is usually required. On the flight in which the accident occurred, the pilot did not contact Denver Center but climbed through 12,500 feet thinking that his 1-2-3-4 code cleared the way. At the same time, the controller at the radar screen that was controlling the Jetstream had the "non-Mode-C filter" on. This meant that any airplane without a Mode-C transponder would not appear on the screen. This was not unusual because this controller was working only traffic above 12,500 feet, all of whom should have had a Mode-C operating. Due to the filter, the Cessna became radar-invisible when he climbed through 12,500 feet. If he had contacted the Denver Center, the controller could have turned off the filter, and the Cessna's location and transponder code would have been displayed on the screen. But the pilot's normalcy had changed. He thought that because he used the same code and daily flew above 12,500 feet without a Mode-C, it was a normal practice to do so. Just then the airplanes collided.

The accident's probable cause was also attributed to the fact that these pilots flew into each other on a day with exceptional visibility. They did not "see and avoid." It appears that the Jetstream had the best opportunity to see the Cessna. The Cessna crossed in front of both the captain's and first officer's forward windshield prior to impact. We cannot know for sure, but it is possible that the Jetstream pilots were not looking out as they should because they had become used to getting conflict alerts from the controllers. Often pilots flying straight and level on IFR clearances will assume separation is guaranteed by the clearance itself and forget to look out when in VFR conditions. They rely heavily on TCAS or controllers to give them heads-up. In this case there was neither TCAS on the Jetstream or a conflict alert from the controller because the controller could not see the Cessna either.

The SOP was once to contact a controller to get radar identified, get a transponder code, and get a rule waiver if needed. The SOP was once to issue random transponder codes, not routine ones. The SOP was once to maintain a visual viligance outside the airplane and not to trust controllers completely for separation when in VFR conditions. But somewhere these SOPs slipped. This certainly was not the first time a pilot did not call in. It was not the first time an "in-house" controller routine evolved. It was not the first time pilots on an IFR clearance did not look. On those other occasions, however, an accident did not take place. Everyone got away with it, and therefore it started to become routine. What was once normal was no longer normal. What was considered normal slipped and evolved into a new normal that was less safe. Then one day these factors came together over Loveland, Colorado.

You can see that calling this simple pilot error is an insult to what actually happened and in a way is a cover up. Writing this off as a simple mistake hides the real reasons. This is an example of a culmination error. The forces that worked together to create this accident developed gradually over time. To get to the bottom of what really caused this accident would require a deep examination.

Not replacing a pitot tube cover is not a terrible offense. But it's not the failure to replace the cover, it is the attitude the pilot developed that made it normal not to cover the tube that is dangerous. That attitude starts the creep. "It's okay; we never use those covers anyway" displays an attitude that stands at the top of a slippery slope. Offenses a little farther down the slope might include logging flight time that did not actually occur to maintain a record of currency. Another would be signing off a maintenance repair when in reality the fix was put off until the next major inspection. When these falsehoods start to become routine and then normal, accidents are sure to follow.

The reason for examining an accident is ultimately to make recommendations that will prevent a similar accident from ever happening again. This brings us back to the question: What can we do to train-out the next accident? One answer is to fight creeping normalcy. This book will expose some problems in the way we train pilots that can allow and actually encourage culmination accidents to occur. It starts with understanding the complete responsibility of being pilot in command.

Acting as pilot in command

The responsibility of being pilot in command (PIC) is a concept, and this book is about that concept. But this idea is often overshadowed by arguments over the legal aspects of "acting" as PIC. So this chapter concludes with an obligatory look at what the regulations say about being PIC.

The regulations mention the phrase "pilot in command" several times in different contexts. Some of these references even seem to contradict each other. Part of the confusion exists because there is a difference between "acting" as PIC and "logging" PIC time.

Pilots must keep records of their experiences, usually in a pilot log-book. Technically pilots need only keep records that prove that they are "current" to fly. They must show that they have had a flight review within the past 24 months, they must have had three takeoff and landings within the past 90 days, and so forth. When a pilot writes the flight time for any flight in the logbook, he or she must decide if that time is PIC time or not. The "logging" of PIC time falls under regulation FAR 61.51. This regulation says that a pilot may log PIC time for any flight in which he or she is serving as PIC or when serving as PIC where more than one pilot is required. So far this makes sense. The regulation is just saying that if you "serve" as a PIC, you can later log that flight as PIC time. But FAR 61.51 goes on to say that even if a pilot is not acting as the flight's PIC, there are circumstances when he or she can log that time as PIC anyway. It seems like a misrepresentation. How can someone who was really not in command later say he or she was in a logbook? The FAA has allowed this confusion to continue in an effort to help pilots build PIC time.

FAR 61.51 says that even if pilots are not acting as the flight's PIC, they can still log PIC time for the portion of the flight when they were the "sole manipulator of the controls" in an aircraft for which they are rated. So let's say that you and I go up in a single-engine airplane. We both are pilots, each rated in single-engine airplanes. During the flight I am responsible for and the final authority as to the operation of the flight, but for 10 minutes during the flight you operated the controls. When we get back down, you can log PIC time for those 10 minutes.

FAR 61.51 gives additional privileges to flight instructors. Flight instructors may log PIC time whenever they provide flight instruction, even if they never touch the controls. So when a pilot and a flight instructor go up in an airplane, both could log 100 percent of the flight as PIC time as long as the instructor was providing instruction to the pilot toward a future certificate or rating.

This distinction between logging only "sole manipulation time" as PIC time and "flight instructor time" as PIC time once cost two pilots their certificates. Two pilots were discovered to have identical logbook entries for over 200 flights. The two pilots were co-owners of an Apache. Whenever they flew together, they both logged PIC time even though only one person at a time can be the sole manipulator of the controls. The FAA took enforcement action against both pilots.

When the case came up, the pilots claimed that the flights were actually instruction flights and submitted "corrected" versions of their logbooks. In the corrected logbook the pilot who was also an instructor had logged all the flights as PIC time because he claimed he had been acting as an instructor. The other pilot, who was not an instructor, logged all the flights and PIC time and as dual instruction received. The federal judge did not buy the "corrected" logbooks and ruled that their original intent was to manufacture PIC time that was not allowed under the rules. The judge revoked the pilot certificates of both pilots. The pilots sought to reduce the penalty to a 90-day suspension on appeal to the National Transportation Safety Board (NTSB) but the board affirmed the revocation (Administrator vs. Crow, et al. NTSB order number EA-4008).

In another twist the regulation says that instructors can log PIC time even when they did not act as the PIC on the flight. When an instructor takes off with a student pilot, the instructor must wear two hats. The instructor is simultaneously acting as the PIC and as the flight instructor. The instructor must be the "acting" PIC because the student pilot has not yet passed a checkride and become rated in the airplane, so the student cannot be the PIC. If the student pilot cannot be the PIC, the instructor must because every flight must have at least one PIC. But what about when an instructor goes up with a pilot who is already rated in the aircraft? Is it possible for a pilot to be the PIC while receiving instruction? Yes. When an instructor goes up with a current private pilot to show that pilot advanced maneuvers, the instructor only wears one hat. The private pilot can be the acting PIC and the instructor provides instruction. Because the instructor is not acting as the PIC, he or she need not meet the qualification to be a PIC. This means that an instructor who, for instance, has no medical certificate, can still give flight instruction as long as the person receiving the instruction is acting as the PIC. In this case the instructor could not fly alone because he or she did not qualify as the PIC but could ride along as an instructor. Can the instructor charge a fee for the instruction in this situation? Yes. The fee being charged would be for "instructor services," not for acting as the PIC.

The regulations offer one more contradiction. FAR 61.51 clearly says that you cannot act as the PIC of an aircraft until you are rated in that aircraft. *Rated* means that the pilot has the appropriate category, class, and type (if required) privileges on his or her pilot certificate for the aircraft being operated. This holds true except on checkrides. FAR 61.47 says that pilot examiners cannot be the PIC during the practical test. So on a private pilot checkride, something unusual happens. On the checkride flight the examiner is not PIC and the student pilot is not yet rated because he or she has not passed the checkride yet. If neither pilot can be PIC, how do they legally get off the ground? The FAA, to protect pilot examiners, looks the other way. It allows the nonrated student pilot to act as the PIC for the purposes of the flight test.

You can see why there are so many differences of opinion about PIC time. People often confuse acting as the final authority on a flight and logging PIC time for the flight. This confusion can lead to larger problems. An FAA inspector arrived on the scene of an accident once where the airplane had run off the side of the runway and hit some trees. There were two people on board the airplane, and both were rated pilots. The inspector asked, "Which one of you was the pilot in command here?" Each pilot pointed at the other.

On any given flight, how is the acting PIC determined? Who really is the final authority when two pilots fly together? Is it the pilot sitting in the left seat, the pilot with the higher certificate, the pilot paying for the flight, the pilot with the greater flight time, the pilot logging PIC time, or the pilot who was both "acting" as PIC and logging PIC time? What if one of the pilots was also a flight instructor? What if it had been a student pilot flying solo; isn't a student pilot's flight instructor the final authority? You can see that the list of PIC questions goes on and on. Shouldn't the pilots work all this out prior to flight?

One more regulation speaks to this issue. FAR 1.1 offers yet another definition of PIC. *Pilot in command* means the person who:

1 Has final authority and responsibility for the operation and safety of the flight

2 Has been designated as PIC before or during the flight

3 Holds the appropriate category, class, and type rating, if appropriate, for the conduct of the flight

Read the second part again: The pilot in command means the person who has been designated as PIC before or during the flight. This says that before each flight when more than one pilot will be aboard, a decision must be made. It should be clear who will act as the flight's PIC. It should be clear who, in addition to the pilot acting as PIC, will also be logging PIC time. It seems logical that these decisions be made before takeoff, but the regulation does also say that the decision could take place "during the flight." When I read that the first time, I pictured this scene: Two pilots flying along enjoying themselves when the engine quits. As the airplane glides back down, the pilot in the left seat says "I'd like to designate you as the PIC now." The pilot in the right seat says, "Oh no! You are the designated PIC here!" The two fight over it without paying much attention to the airplane situation. The airplane stalls and spins in.

Designating the acting PIC prior to flight is a great idea. Today, in airline operations the same thing happens. On an airline flight deck there are two pilots. On any given leg of a trip there will be a pilot flying and a pilot not flying the airplane. The pilot flying the airplane has specific SOP duties. The pilot not flying has different SOP duties. The captain of the airplane has final authority, but the pilot flying/pilot not flying concept makes the two pilots equal partners, and this improves crew resource management. General aviation pilots should benefit from this example. Designate the acting PIC before takeoff and make that the pilot flying the airplane. If the circumstances of the flight allow, also designate which pilot(s) can log PIC time before takeoff.

Practical advice

Even though in certain situations the FAA allows two people to log PIC time simultaneously, employers do not often recognize this time as valid. If you are building flight time in hopes of landing an airline position, you should know that the airlines do not count your "logged" PIC time toward their minimum requirements. The airlines will have you subtract from your logged PIC time all the time that was with an instructor. In other words, the airlines do not play the FAA's acting/logging game.

Let's say that you are a multiengine pilot looking to increase your total time. One day you get the chance to fly on the "dead-head" part 91 portion of an air taxi flight in a Cessna 421. You ride along on the trip in the right seat and the pilot even lets you manipulate the controls for awhile. On this flight the company pilot is the acting PIC, but later you log the time as multiengine PIC time because you were rated for multiengine and manipulated the controls. Time goes by and later you get called in to interview for a pilot position. A keen interviewer will notice that "way back when" you were the PIC on a Cessna 421 flight. But this Cessna 421 flight has come from nowhere.

There is no Cessna 421 flight training logged anywhere, and he or she knows that you can't just walk out and fly a Cessna 421 as PIC with no previous training on that airplane. The interviewer will immediately see through your scheme and realize that what you are presenting is in fact only logged time. He or she will know who was really in charge on that flight and it wasn't you.

Representing logged flights as acting flights will be transparent to the interviewer. If you have relied on these flights to meet the company's minimum flight requirements, you are going to have a short interview and be back on the street. You will have wasted the interviewer's time and probably removed any future chance with that company.

A much better scenario with the Cessna 421 flight could have taken place if the company pilot had also been a multiengine flight instructor. In that case you could have logged PIC as instruction received. This will not do anything to help meet minimum flight times, but it might be impressive to the interviewer that you had sought out opportunities to get additional multiengine instruction. This at least would not be a misrepresentation of what really happened.

All too often we get caught up worrying about the legal side of PIC. We debate the FARs to the point that the issue of logging PIC time or acting as PIC in the legal sense shifts the meaning of pilot in command to a legal definition. The legal definition has its place, but being the PIC is much more and much different.

Taking the responsibility of PIC means taking responsibility for combining information and situations to make timely decisions that produce safe outcomes. The PIC is not only the person who manipulates the controls of a machine, but the one who controls all the human elements, seeks knowledge on the ground and in the air, and can be consistently counted on to do the right thing. The concept of pilot in command is one of those things that is hard to define, but you know it when you see it.

2

What Are the Airlines Doing that We Should Be Doing?

The Air Safety Foundation issues the Joseph T. Nall report each year. The Nall Report is a publication of aircraft accidents within the general aviation industry. Year after year these accidents form trends that can be tracked. The general aviation accident rate was 10.3 accidents for every 100,000 flight hours in 1996. That number was up from 7.8 accidents for every 100,000 flight hours in 1990. The general aviation accident trend is on the rise. During this same time period, accident rates within the airlines have been lower and on the decline. In 1990 airline accidents stood at 1.5 accidents per 100,000 flight hours, and in 1996 it was just less than 1.0 accident per 100,000 hours of flight.

The National Transportation Safety Board classified 65 percent of the general aviation accidents as pilot error accidents. The Nall Report estimates that when accidents that were first categorized with an "unknown" cause are completely investigated, the pilot-error category increases to approximately 80%. This means that pilot decision making and/or pilot error is by far the greatest cause of accidents.

Is pilot error also a large causal factor within the airline industry? The *Air Line Pilots Association* (ALPA) asserts that accidents that are caused by pilot error of airline pilots is declining. The ALPA also utilizes NTSB accident statistics and claims that flight crew error has not been a factor in the majority of fatal accidents involving major air carriers during the past 10 years. Gerard M. Bruggink is a former director of accident investigation at the National Transportation Safety Board. In a 1997 article in ALPA's journal, *Airline Pilot,* Bruggink points out that from 1977 to 1986 there were 18 fatal airline accidents. Of these 18, only 5 were "beyond crew control." In other words, 13 accidents, or approximately 72 percent, were within flight crew control, and the fact that an accident occurred could be attributed to pilot error. From 1987 to 1996, however, 15 of 24 fatal airline accidents were beyond flight crew control, or approximately 37 percent were due to pilot error.

What explanation does ALPA give for the reduction in pilot error accidents? Again Bruggink says, "Although we have no proven explanation for the changing accident pattern, two factors come to mind: (1) the growing emphasis on better team performance of flight crews that began in the 1980s and (2) the gradual but consistent shift to regional carriers of short-haul passenger operations with large numbers of takeoffs and landings."

Bruggink is giving partial credit for safer skies to what he calls "better team performance of flight crews." This is a clear reference to the field of crew resource management (CRM) and pilot decision making. This assertion claims that CRM programs and better awareness of decision making inside the airline industry have produced a 10-year reduction in pilot error accidents.

This means that there are opposing trends within the aviation industry as a whole. The general aviation pilot error accident rate is increasing, and the airline pilot error accident rate is decreasing. What possible variables could be at work here to make these trends oppose? The most obvious answer lies with the experience of the pilots in each group. We are not comparing apples to apples when we compare the average general aviation pilot to the average airline pilot. Most likely the airline pilot was once a general aviation pilot who, because of his or her gain in experience, eventually qualified to be an airline pilot. So the difference in experience is a given, but airline pilots do not stop training once they become airline pilots, so a logical next place to look would be in areas where differences exist with instruction and training. One big difference is found in the area of "decision training."

The airlines have become the proving ground and therefore the beneficiaries of a new field of decision-making theory and apparently have a better safety record to show for it. Can the theories that are being applied in the airline industry be adapted into the general aviation segment, with a reversal of the pilot error accident trend being the end result? This question requires a thorough look at this new field that has evolved over the years into what is now called naturalistic decision making.

Static versus dynamic decision making

"Decision theory as a topic in psychology seems to come in two distinct and almost noninteracting chunks, which I have named static decision theory and dynamic decision theory," concluded Dr. Ward Edwards, of the University of Michigan Engineering Psychology Laboratory, in 1962. He made an early distinction in how decisions are made and the situations that surround decisions. Static decision theory conceives of a decision maker who is confronted by a well-defined set of possible courses of action. Associated with each course of action is a payoff. The decision maker chooses and executes one of the courses of action, receives the payoff, and that is that. The decision maker can weigh the pros and cons of any course of action without time constraints. Real-life decisions such as choosing a spouse, taking a particular job, or buying a house might be examples of static decisions. In life we all face these types of decisions, and the conventional wisdom is to take time to think them out. My father would always say, "You need to sleep on that." Parents especially coach you not to "rush into anything."

Static decisions also seem to be "one-shot" decisions. You make the decision and then go on with your life. In its simplest form there is no follow-up or second decision where lessons learned from the first decision are applied.

On the other hand dynamic decision makers are conceived of as making a sequence of decisions. Earlier decisions, in general, produce both payoffs and information. This information may or may not be relevant to the improvement of later decisions. In dynamic situations, a new complication that is not found in the static situation arises. The environment in which the decision is set may be changing. This change may be the product of a decision that has already been made, or the change could be completely independent. This type of decision making has been called *real-life, real-time,* and *real-world* decisions.

In 1990, Berndt Brehmer added a new element to the mix with his concept of real-time in his article, "Strategies in Real-Time Dynamic Decision Making." Brehmer claims that it is not sufficient to make the correct decisions and make them in the correct order, but that the decision must also be made at the correct moment in time.

Brehmer illustrates his theory by using the scenario of a fire chief charged with the task of fighting a forest fire. The fire chief has at his disposal several fire-fighting units (FFUs). The chief must make decisions about where each FFU should be sent, what supplies each FFU has, the time it takes an FFU to reach the fire, and the speed at which the fire is spreading. All these decisions affect each other. If an FFU is sent to one location and then it is determined that it should now move to another location, time will be lost as it relocates. The time lost could trigger yet another decision if the fire is moving or if the wind shifts. It is clear that decisions made in this situation will be different from decisions made in a controlled, calm, nontime-sensitive situation.

Put differently, dynamic decision making in real time requires the decision maker to cope with processes rather than events. The act of decision making then becomes a form of control over the process, rather than a choice between alternatives. Decision making becomes a control issue, and who is in control when this is applied to airplanes? The pilot in command.

Naturalistic decision making

Researchers Judith Orasanu and Terry Connolly modified the concept once again in their 1995 article, "The Reinvention of Decision Making." Orasanu and Connolly take the field from dynamic decision making to what they call naturalistic decision making (NDM). It is called naturalistic because these decisions are faced in natural or real-life settings. In real life, decisions are not always clear-cut. There is not always a well-defined choice like chocolate or vanilla. In the real world choices are very complicated. Orasanu and Connolly put forth several factors that they believe characterize decision making in naturalistic settings:

1 *Ill-structured problems.* Real decision problems rarely present themselves in the neat, complete form that easily fits into a decision-making model. The decision maker may not even recognize that the situation is one in which a decision is required.

2 *Uncertain dynamic environments.* Naturalistic decision making typically takes place in a world of incomplete and imperfect information. The decision maker has information about some part of the problem but not about others. Also the task is likely to be dynamic—the environment may change quickly, within the frame of the required decision.

3 *Shifting, ill-defined, or competing goals.* Outside the laboratory, it is rare for a decision to be dominated by a single, well-understood goal or value. In the real world the decision maker will be driven by multiple purposes, not all of them clear and some of which oppose each other.

4 *Action/feedback loops.* The traditional decision models are concerned with an event, a point in time in which a single decisive action is chosen. In NDM, in contrast, it is much more common to find an entire series of events, a string of actions over time that are intended to deal with the problem, find out more about it, or both.

5 *Time stress.* An obvious feature of many NDM settings is that decisions are made under significant time pressure. Time pressures have several implications. First, decision makers in these settings will often experience high levels of personal stress, with the potential for exhaustion and loss of vigilance. Second, due to the lack of time to "think out the problem," decision makers use less-complicated reasoning strategies. Decision strategies that demand deliberation over time are simply not feasible. It seems unlikely that reflective thought is the key to successful action.

6 *High stakes.* Examples of NDM settings involve decisions that may affect the loss of life or property, the continuation of a career, or the survival of a business. Obviously, many everyday decisions do not have stakes that are this high, but NDM concentrates on the settings where the participants are directly involved with an outcome.

7 *Multiple players.* Many of the problems of interest to NDM researchers involve not a single decision maker, but several, perhaps many, individuals who are actively involved in one role or another.

As a pilot, do these characteristic situations sound familiar? I believe they are very familiar. In fact I believe that these characteristics define the environment that the pilot in command works within. The pilot/decision maker certainly must make decisions, sometimes with limited information. The pilot in command must make choices under time pressure and with very high stakes. The pilot is obviously involved with the outcome or consequences of his or her decisions. You know the old saying: "Doctors bury their mistakes, but pilots are buried with them."

How decisions are made in stressful environments raises questions in many areas, including flight training. Can a person be a safe pilot before he or she reaches a level of experience that allows for good decision making? Can any training intervention substitute for experience in the meantime?

It is clear that humans can get better at something the more they practice. So why not practice these high-stakes situations and get accustomed to making decisions in these environments? The answer is that many pilots do, but unfortunately these pilots are not general aviation pilots. Airline and military pilots routinely train in real-world environments, and this is one of the biggest differences between what they do and what we do.

The airline accident rate is going down, whereas the General Aviation accident rate is going up. What are they doing that we are not doing? What are they doing that we should be doing? There are two answers to these questions. First, the airline pilots train more often than general aviation pilots. Airline pilots do more work on proficiency than a biannual flight review can possibly provide. Second, when they train they use real-world situations and therefore are practicing their decision making in action.

So if the answer is so easy, why aren't we changing? The reasons that no change has taken place are cost, equipment, and training methods. (1) It costs more money to train more often. Several years ago the FAA wanted to require an annual rather than a biannual flight review, but this proposal was turned back by general aviation pilot groups as being too costly. (2) General aviation pilots do not have access to sophisticated, highest quality, simulation equipment, and even if we did have access, we could not afford to use it. (3) Flight instructors tend to give flight training much like they received it. Our training methods have been "maneuver-based" for many years, and that is the way it remains, handed down from one certified flight instructor (CFI) to another. The FAA practical tests rely heavily on maneuvers and less on decisions, so the CFIs are not completely to blame. After all it is much easier to grade a short field landing than a person's decision capability.

The deterrents of cost, equipment, and training methods are formidable, but if we are to reverse and lower the accident trends of general aviation, we must attack them. The airlines and military are not perfect. They do not hold the exact solution to all of our problems, but they have fewer accidents, so we must take a look at what they are doing. It is a cop-out to simply say that they have fewer accidents because of their greater experience. That is giving up without a fight. Are the lives of airline passengers more valuable than the lives of general aviation passengers? Of course not. So we must observe what we in general aviation are doing now, and be willing to make improvements, even if that means changing attitudes that have been around a long time.

3

General Aviation Left Behind

When I was in college in the late 1970s, I took a field trip in one of my classes to the administration building to see the university's computer. We went down into the basement, and spread out behind a glass wall was a row of machines, each refrigerator size and each with two reels of tape whirling away. The computer room's floor was raised so that all the people inside the room were at least 1 foot higher than those of us on the outside. The students in my tour group were not allowed to actually enter the computer room, and when we left, I did not know what those machines did or how they were helping me.

Imagine the difference today. At that same university rather than one computer, now there are thousands. Not only are there thousands of computers, the students all know how to use them, and any one of them is more powerful than that one basement-sized computer of the 1970s. As computer technology has exploded, many industries have jumped on the bandwagon and uses include flight simulation. In the time since the 1970s, flight simulation has gone from a GAT (general aviation trainer) to full-size, holographic, multiple-axis motion marvels. The flight simulators of today are so real in every respect that it has become common place to do 100 percent of the required flight training in them. Today when a recently hired airline pilot flies the actual airplane for the very first time, that airplane is full of paying passengers on their way to Chicago or someplace else.

These fantastic computer/simulators also offer the best opportunities yet seen to test pilot decision making. Complete start-to-finish flight scenarios are possible now, and researchers have taken advantage of it. This all sounds like wonderful news and it is, except for one problem: General aviation has been left out. With each improvement in the computer/simulator technology the price has increased. Over a 25-year period, general aviation simply was priced out of the market.

Researchers, seeking a device that would be closest to the real thing, followed the technology up the scale. Because the airlines and military were the only ones that could afford these devices, the researchers were drawn to airline training centers and military bases and away from general aviation airports and its pilots. In the last 25 years there has been some real ground-breaking work done. Crew resource management, once thought of as "charm school," has come of age as a life-saving tool. Decision theory was tested and training strategies were altered as a result. Required pilot skills moved away from the *Right Stuff* to the ability to utilize all resources while remaining in command. It was a golden age of discovery. But during this time there was not a single study using the new generation of computer/simulators that was aimed at general aviation. The research and all its benefits left general aviation back in the 1970s. Is it any wonder that our accident rates are higher?

While the airlines and military were training pilots to make decisions in critical situations, we were still teaching and learning how to fly S turns and chandelles. There was nothing wrong with learning the lessons that are provided by maneuvers like S turns and chandelles, but on that proverbial "dark and stormy" night I would feel better with a decision maker, not simply a maneuver maker, at the controls and in command.

The airlines use a form of training called line-oriented flight training, or LOFT. *Line* refers to the *flight line,* in other words, what goes on in the real world. In a LOFT scenario a pilot and crew are in the simulator but fly through an actual flight from Seattle to Los Angeles, as an example. Along the way the simulated flight will encounter many decision-prompting situations called *event sets.* This exposes the pilot to real-world events and places them in decision situations. It is this type of training that has yielded declining accident rates for the airlines, so why not use this type of scenario-based training with general aviation pilots?

Single pilot rather than a crew of pilots

The work combining computer and simulators and decision training inside the airline companies involves flight crews. A flight crew is more than one pilot and is supposed to work together to provide a safe and efficient flight. But general aviation does not rely on flight crews. Pilots are trained in general aviation to act alone.

Single-pilot operation is a way of life in general aviation. Flying as the single pilot in a high-workload environment provides different challenges. The National Business Aircraft Association (NBAA) has focused on the problem. The NBAA is an industry group that gives a voice to the nation's pilots who fly for corporations. The pilots who fly for corporations that own airplanes are not unionized and depend on the NBAA to provide guidance on many issues. One issue that the NBAA addresses is the use of a single pilot to fly in instrument conditions. "The risk of an accident or incident in operations with a single pilot operating under IFR is significantly higher than operations with two pilots under IFR. The NBAA realizes that some companies need to operate with a single pilot in operation under IFR; however, these operations place considerable workload on the pilot. The NBAA therefore recommends that the following precautions be taken in any single-pilot operation under IFR: (1) An autopilot with at least a heading hold feature should be used, (2) pilots should have 20 hours of instrument flight experience in the particular airplane flown, (3) pilots should perform detailed preflight planning, and (4) pilots should receive an instrument competency check every 6 months" (NBAA Management Guide, 1997).

The NBAA guide says that the risk of an accident is "significantly higher" when operating IFR with only one pilot. But the NBAA also realizes that single-pilot operations are an economic fact of life, so they offer their list of recommendations in hopes of reducing the risk. These NBAA recommendations are just that, recommended not required. Furthermore these guidelines are not even seen by the thousands of general aviation pilots who fly single-pilot IFR and are not NBAA members.

There has been little or no research aimed at general aviation pilots. There are few guidelines for general aviation pilots to follow. They most often fly as the single pilot and have never been trained in or exposed to crew concepts. Therefore safety strategies designed for crews pass them by. If accident statistics are ever to drastically improve, these facts must change.

The use of flight simulation

There is a debate about the best method to observe pilots and their decision-making tendencies. There seems to be four ways to do it: the classroom, computer-assisted instruction, flight simulation, and flight in an actual airplane. The classroom method usually involves a retrospective look at decisions made by pilots who were previously in accidents. The method involves recreating the decision sequence that led up to the accident. This is the "learn from the mistakes of others" method, and it is better than nothing, but it is not very real to the pilot. There are several good computer programs that test pilot decisions. Still, I do not believe that sitting at a computer provides the same situation urgency that an airplane does, and therefore the decisions made are too "laboratory" produced. Of course, you could go up in an airplane and place the trainee in the same hazard that killed the other pilot, but that is too real. The safe alternative appears to be flight simulation. To me flight simulation strikes the compromise between realism and safety. Although the participants know that it is a simulation and that their life is truly not in danger, the sweat-through clothes, the body language, and the stress of a simulator session tend to suggest that it is real enough to see actual pilot decision behavior.

The flight simulator and flight training device distinction

Some terminology problems have developed over the years in the world of ground simulation. The term *flight simulator* has become misunderstood. Essentially an airplane "simulator" must be a full-sized exact replica of a specific type, make, model, and series airplane cockpit. It must have computer software programs that are necessary to represent the airplane in ground and flight operations. It must have a visual system that provides an out-of-cockpit view and a "force (motion) cueing" system that at least provides cues around the three axis of airplane control (FAA Advisory Circular-120-45A, 1992). These are sometimes referred to as "full-motion" simulators. A simulator that has all the requirements just described has a price tag starting at $8 million. This price makes the use of flight simulators prohibitive outside of military and airline training centers. General aviation pilots are more likely to use what is now being called a flight training device (FTD).

An FTD is also a full-sized replica of an airplane that uses computer software to present ground and flight situations, but it does not move. The FTD's layout can be inside an enclosed cockpit or arranged as an open flight deck. This device can be a generic trainer.

In other words, it does not have to duplicate any specific airplane, and visuals are not required. This category of training device has gone by many names in the past and this has led to some confusion. Cockpit procedures trainers (CPT), cockpit systems simulators (CSS), ground trainers (GT), and fixed base simulators (FBS) all now fit under the FTD category.

Before any machine is classified and can be used in the training of pilots, it must be approved by the FAA. A principal operations inspector (POI) from the local FAA office must inspect the machine. A machine is not a simulator or FTD until the FAA says that it is.

Computers alone will not do it. The benefits to general aviation from research may still be years away. So to lower accident rates and improve the overall safety of general aviation, those inside general aviation must take the lead. We must move from maneuver-based training to mission-based training on our own. Fortunately we have some role models. General aviation includes many pilots, from novice to expert. We know that experts are safer, so why not pattern the decision behavior of experts to make improvements?

4

Expert Pilots

Hundreds of pilots who work in general aviation every day could be considered experts. These pilots fly everything from glass cockpit corporate business jets to crop dusters. They earn their living flying transoceanic flights, meeting the schedule of "on-demand" passengers, hauling cargo, and giving flight instruction. The difference between an expert pilot and a novice pilot is simple: experience and the ability to make decisions. Experience and decision making go together, but it is not clear which comes first. Pilots who have had the benefit of many flight hours may have seen situations during those flight hours that help them make decisions. In other words, their experience guides their decisions. Or it may be that pilots who can make good decisions in the first place remain on the job longer and live longer. They have the longevity that allows them to gain the experience. It probably is a combination of the two. Either way there is a definite difference between novice pilot behavior and expert pilot behavior.

If there really is an observable and measurable difference between expert pilot performance and novice pilot performance, can there also be teaching strategies developed that would help novices act more like experts? If such a strategy were developed, it would have to start with a definition of just what expert pilot behavior is. So what do expert pilots do that is different and worth copying?

Characteristics of expert performance

Expert pilots are able to anticipate and prepare far more than the novice pilots. Experts can fly with ease, never coming near a mental saturation point. This leaves them with the mental capacity available to think ahead and plan for upcoming events. Experts never seem to be in a hurry, yet they are always doing something. They never let a free moment go without planning something or doing something that will help them out later. Expert pilots do all the extras and little things that make the job easier.

In any flight procedure there are several task layers. There are tasks that absolutely have to be done if the flight procedure is even possible. An example of this would be tuning in a navigation radio to a frequency that is used for an instrument approach. Without that frequency the pilot cannot know where to go, so tuning that frequency is an absolute necessity.

Then there are tasks on a slightly higher level that, although not absolutely required, make the procedure run smoothly. An example is prereading the missed-approach instructions so that when the time comes for the missed-approach procedure to be executed, experts calmly add power and begin the procedure without immediate reference to any chart. Nonexpert pilots rarely are that prepared at the missed-approach point. They often fumble around looking for the proper chart while the airplane is somewhat out of control.

The third task layer involves situation awareness management. One example of the expert at work is dialing in an additional navigation radio frequency on a second radio, even though this second radio is not required for the flight procedure at hand. Experts use it anyway to more clearly determine their position. With this knowledge the experts are aware of their relative position throughout the procedure and are able to call on this knowledge. At times they will turn with a tighter radius to make a smooth course intercept. The only way the expert could have known that a tighter radius was called for was having knowledge of relative position. With the course intercept made smoothly, the approach procedure begins under control and no time is wasted passing through the course and attempting to reintercept from the other side. Experts are constantly and predictably completing these extra third-level tasks. The result is that the procedure appears effortless and everything is under control. One commonality that all experts have is the ability to physically fly the airplane without using up all their mental energy. These pilots are able to hold altitude and heading when that is required and still plan ahead. When the physical workload increases, such as a turn or descent, a descending turn, or a course intercept, they are capable of keeping pace both with the physical task of manipulating the airplane controls and also with the mental tasks. They do not miss radio calls. They are assertive and clear with radio transmissions. They do not miss altitude changes. They reduced speed when they should. They are in command.

Experts never seem to "get behind the airplane," but this is no accident. They are always planning ahead. They are always doing something that although it is not actually mandatory at the time will pay off soon thereafter. Here is a short list of some preparations that expert pilots make:

1 Experts set a VOR (very high frequency omnidirectional range) radio to an outbound course before arriving at the station.
2 Expert pilots ask for an updated wind report when turning on the final approach course or on short final when landing.
3 Experts tune in a backup navigation frequency on the second radio.
4 Before flying into an air traffic control sector, experts observe the communications radio frequency for that upcoming sector and tune in that frequency on the second communications radio when the workload is light. When the time comes to switch to the new frequency, it only will take the flip of a switch and the workload at that point is reduced.
5 When faced with instrument approach weather decisions, experts ask for weather reports from many surrounding airports before arriving at a final decision.
6 Experts under a heavy workload circumstance will solicit information from the controller. During "crunch" time, experts might ask about the length of a runway or the tower frequency or the direction of traffic flow, rather than attempting to look this information up in a directory or chart book.
7 Experts, when faced with unknown circumstances, make backup plans. If the pilots face an instrument approach with weather at or near the minimums, experts make plans in advance for the possibility of a missed approach or a circle to land or even a tailwind landing.
8 Experts use all their resources, which includes nonpilot passengers. These pilots have nonpilot passengers look for other air traffic, hold chart books, flip pages, and any other task that would be helpful during high workload events.
9 Experts use backup radios to listen to prerecorded weather broadcasts (Airport Terminal Information Service, or ATIS).
10 Expert pilots anticipate station passages so that they can start and end timed segments of the flight properly.

11 Expert pilots make and take the time to listen to the Morse code identifying broadcast of navigation radios.

12 Experts update altimeter settings as the flight progresses. They double-check for the proper setting before and during instrument approaches to ensure that the proper minimum descent altitude or decision height is reached.

13 When expert pilots are unable to land due to low clouds at one airport, they ask the controller if other airplanes have recently landed on an instrument approach at a nearby airport. The logic is that if other pilots are landing, it would also be possible for them to land as well. This helps make alternate airport decisions.

14 Expert pilots ask for additional information from other pilots who have flown the course ahead of them.

15 Expert pilots do not hesitate to discuss problems that arise with air traffic controllers.

16 When airplane malfunctions occur, expert pilots take appropriate internal action and advise controllers on their situation and what impact the malfunction has on the remainder of the flight.

Experts do very little talking during the flight except to air traffic controllers. They very often talk to themselves, however. These internal conversations can be characterized as reminders and questions to themselves. Some common examples of these are:

I've got that set up in number 2.
I'm ready to make the turn outside the marker beacon.
I'm at 9000—1000 feet to go until level off.
Can I get a DME (distance measurement equipment) reading off of Nashville from here?

These are not addressed to the controllers but rather are audible thoughts. When the expert pilots do address controllers, they often have suggestions for the controllers. On the surface this might seem somewhat backward. Most of the time you think of controllers telling the pilots what to do, but experts never wait for a controller to come up with a plan. They are ahead of the controller:

Can I get a 320 heading now?
I can accept a tight turn-on to the localizer.
Say again the ceiling at Atlanta.
Confirm you want a left turn; my chart indicates a right turn on the missed procedure.
How much longer will you need me on this heading?

These pilot suggestions are usually made professionally but forcefully. The fact that the pilot knows enough about the situation to be making suggestions to the controllers is evidence of his or her awareness.

Experts are aware of what is taking place around them. They are never caught off guard. They do not miss any important clues or information coming from inside or outside the airplane. They take on each challenge and work it through to a logical conclusion. Whenever a solution is not immediately found to a problem, an alternative is decided upon. Expert pilots are expert troubleshooters. They know their airplane's systems and can diagnose problems.

There really is a big gap between experts and novices. I have discovered that the problem is bigger than I once anticipated. The challenge for me then and for every flight instructor today is to figure out ways to help pilots act more like experts, even though they have low flight time and little experience. It would seem logical that if we want to act like experts, we should just do what they do. So, we need a map to show us the way to expert pilot behavior. To help, I designed a model to help novice pilots remember to do what experts do. It is called the ASAP model.

5

The ASAP Decision Model

Most people understand the letters ASAP to mean "as soon as possible," and that metaphor of urgency applies to pilots who must make decisions under stress, but the letters ASAP applied to pilots stand for anticipate, situation awareness, action, and preparation. ASAP is just a memory jog. It is something to help pilots concentrate and zero in on the problem so they can more easily solve it.

Researchers Kaempf and Orasanu addressed the use of models in their 1997 article, "Current and Future Applications of Naturalistic Decision Making in Aviation." They concluded that under conditions of time pressure and ambiguity, decision makers need help to determine what is occurring in the environment around them. This is equally true for flight crews, air traffic control, and many other decision makers. Therefore, decision aids, interfaces, and training should provide decision makers with the tools and skills necessary to accurately and quickly make situation assessments. The ASAP model is such a tool.

How many times have you heard a flight instructor tell you to "stay ahead of the airplane"? What does this actually mean and why is it a good thing? "Stay ahead" is a figure of speech of course; you cannot arrive at your destination ahead of the airplane. It means that in your head you are predicting what will take place next and planning for it to happen. Staying ahead of the airplane means that you seldom get caught off guard. When you are ready for the next event, a flight seems to slow down. When you are unready and constantly being surprised or unaware of events, a flight seems to speed up. Remember, experts in flight never appear to be rushed, hurried, or panicked. They are always ahead of the airplane.

There is also a link between being ahead of the airplane and quality decision making. Mica Endsley in a 1995 work on this subject, "Toward a Theory of Situational Awareness in Dynamic Systems," concluded:

> *There is considerable evidence that a person's manner of characterizing a situation will determine the decision process chosen to solve the problem. Situation parameters or context of a problem largely determines the ability of individuals to adopt an effective problem-solving strategy. It is the situation specifics that determine the adoption of an appropriate mental model, that leads to the selection of problem solving strategies. In the absence of an appropriate model, people will often fail to solve a new problem, even though they would have to apply the same logic as that used for a familiar problem.*

This means that people make better decisions when they start out from a position of knowledge and awareness. Therefore to end with a positive decision a pilot must start with awareness. To gain or maintain awareness a pilot needs a model or a reminder to get started. ASAP is such a reminder.

The ASAP model is really just a gimmick to help pilots stay ahead of the airplane because common sense tells us that staying ahead leads to better and safer decisions.

Anticipation

Anticipation (Fig. 5-1) is another way of saying stay ahead or be prepared. But it is easy to say it, harder to actually do it. To get ahead and stay ahead of the airplane, try the following methods.

Put money in the bank

Pilots should put money in the bank. Conventional wisdom about everyday life is that you should have a little savings set aside for a rainy day. In this way when your car breaks down, you can get it fixed and still pay the rent. The savings become a cushion or a buffer against the unknown or unexpected. To the pilot, putting money in the bank means doing all the extra things early. This preparation will act as a buffer against the unknown or unexpected flight event. If you are flying level en route and not much is happening at the time, look on the chart and see what frequency is likely to be your next handoff. Dial that frequency into your second radio so it is ready to

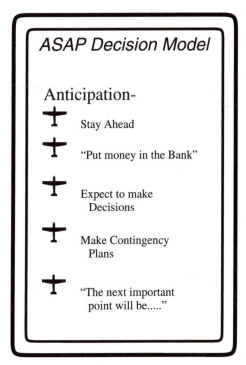

Fig. 5-1 *Anticipation component of ASAP decision model.*

go. If something unexpected should come up (turbulence, traffic conflict, sick passenger, rough engine, equipment malfunction, etc.) before or during the controller handoff, you will have less to do because the frequency will already be waiting for you, and you will have more time to deal with the unexpected problem. You will be very busy approaching a high-density airport, so always have the next frequency ready. This way when the controller points out three other airplanes in your area and asks you to keep your speed up, make a tight base turn, follow the 737, report the airport in sight, and contact the tower, you can throw one switch on the audio panel rather than fumbling for the frequency and dialing it in.

Some VOR approaches do not require DME (distance measurement equipment) but nevertheless have DME positions on the approach chart. Put money in the bank on the VOR approach by setting up the DME early and using it throughout the approach. You can determine your outbound distance, for instance, so that you do not fly outside the 10-mile protected area of the approach. On the way inbound you can get a better idea of how quickly you must descend to get down to minimum descent altitude (MDA) in time to make the airport. If you do not have the DME, you could use a crossing radial from another VOR to mark the outside of the 10-mile protected area and an intermediate point along the final descent when you should be at MDA. There are countless more examples of putting money in the bank. In each case we are doing something when we have time to better protect us for when we do not have time. There are times during any flight that can be downright boring. But the aware pilot knows that these times will not last. Soon the situation will change, and the pilot workload will go from boredom to blowout. The next time you are flying along without much happening, remember to use this time wisely. Think of ways that you can put money in the bank.

Expect to make decisions

Accident reports are filled with circumstances where a pilot was caught off guard by the events of a flight and was simply carried away by them. Many pilots do not understand that making decisions is as common to a flight as takeoff and landing. There are situations on every flight where a decision is called for, and every second that passes without making the decision makes the situation more critical. In these situations some pilots are just unaware that a decision is staring them in the face and screaming "solve me!" There can be a real lack of understanding that decisions are necessary to the safe outcome of the flight. Some pilots assume that controllers would not let them get into too much trouble or that things will just work out because things always seem to work out. These attitudes lead to crashes. Reading some of these accident reports, I envisioned the pilots walking like zombies through a mine field with explosions going off all around them. They walk along unaware of the dangers until it is too late. They miss the warning signs, and they miss opportunities to decide on a course of action that would solve their problems. Some, while completely unaware, happily press on to their own fatal accidents.

The solution? Pilots must realize that making decisions is their job as pilot in command and that there will be countless points of decision on every flight. Pilots must accept this fact and start to expect to make decisions because decision making is normal. When flying along to a destination and when the pilot workload is low, the pilot should be planning for and anticipating the decisions that are bound to present themselves. Which runway is in use at the destination? How should I maneuver to be in a position to enter the pattern? Should I expect to fly an instrument approach at the destination? If I do shoot the approach, should I be planning on a straight-in or circling approach? These are examples of decisions that are inevitable. You will not be able to see perfectly into the future, but you can plan and expect to make these decisions as a routine function of any flight. Understanding that pilots have flown deeper and deeper into trouble because they could not see that decisions were called for, I want to shake them awake. I want to make them snap out of it as if they were under hypnosis and yell "you are PIC; don't just sit there! Make some decision, any decision, it's your job, your duty!"

Decisions are not the product of some abnormal flight situation that a pilot might never face in a career of flying; they are everyday, normal, and routine. Pilots should expect to make decisions on every flight. Pilots must search, like a detective, routing out the hidden decision need that calls for PIC attention.

Make contingency plans

Pilots must never waste any time. Expert pilots are always doing something, even when the workload is low. They may only be holding an altitude and tracking a VOR radial, but they are always thinking and doing something. During these times the pilot should play a little mind game. They should play what-if. Ask yourself as you fly along What if I started to get a rough running engine, what would I do? Are there airports near by that I could divert to? Are there fields below I could land in? The engine will probably not run rough and the flight will probably continue on without any problems, but asking what-if makes you think of a contingency plan just in case. Have you ever heard the saying, always leave yourself an out? This means having a backup plan. You should switch to the backup plan anytime the first plan does not seem like a good idea anymore. But you must have a backup plan waiting in reserve to be able to switch to it. Playing what-if forces you to think up a backup plan.

What if I get to the decision height and cannot see the runway? What if I am told to switch frequencies by air traffic control (ATC) but cannot get anyone to answer on the new frequency? What if my approach clearance gets canceled right here and I am asked to do a holding pattern? What if the glide slope goes out in the middle of an ILS approach? What if I start picking up ice at this altitude? Each of these what-if questions would force the pilot to answer with a contingency plan. Someday one of your what-if questions might come true, but because you have asked them and prepared a response, you have a ready-to-go contingency plan. On that day you will not waste valuable time wondering what you should do. You will quickly decide to switch plans, solve problems, and move on to the next what-if.

The next important point will be...

Practice anticipation during the low workload times by filling in this blank: The next important point will be _____. I have seen this situation played out many times. A controller asks a pilot to fly to a particular fix like a VOR or a nondirectional beacon (NDB) and hold over the station. The pilot turns the airplane and begins to fly toward the station. There may be a crosswind, so the pilot works out an angle that will offset the wind and create a straight course to the fix. Then we fly in silence for the several minutes it takes to get to the fix. All of a sudden, as if it's a big surprise, the station is passed and the pilot now does not know what to do to enter the holding pattern. What kind of hold entry should I use? Do I turn? If so, what heading should I turn to? Which direction should I turn? All these questions come into the pilot's head in a mad rush as the station is passed. Confusion sets in, altitude is lost, radio communications are forgotten, the holding pattern's protected area is flown out of; in short, all hell breaks loose. All that could have easily been avoided if the pilot had used the time traveling to the fix to plan for what was needed after the fix. So to remember to plan ahead, fill in the blank: The next important point will be _____. That point might be passing a holding fix, and that will remind you to plan for the hold entry to follow the fix. The next important point may be an intercept and a descent on an approach. It might be reporting your position flying inbound to a VFR tower. It could be anything, but ask yourself what it is and you will be planning ahead.

Anticipation then means using time wisely, putting money in the bank, expecting to be in decision situations, asking what-if questions, and reminding yourself what important event is coming up next. If pilots do this, they will be better prepared when and if things go wrong. They will be less likely to be caught off guard or to be unaware when something is going on that they should know about. This all implies, of course, that the pilot must keep an active mind during the entire flight. There is no room for the pilot to doze off mentally or to be thinking about nonflight problems. Mental alertness leads to situation awareness.

Situation awareness

Situation awareness (Fig. 5-2) probably has as many meanings as there are people who wish to define it. The essence of it would be the pilot's ability to understand the situation at hand and to utilize all available resources to maintain that understanding. Joel Smith, Fleet Training Manager for the 747-400 at Northwest Airlines (and former student of mine), defines it by saying, "wherever you are—be there!" Pilots must learn and practice this art of awareness. The ASAP model offers the following suggestions to gain and maintain awareness.

Develop the mental autopilot

Some students get impatient with the first lessons toward the instrument rating. They get bored with hood work like climbing turns, the vertical S, timed turns, constant rate descents, and standard rate turns. They want to get on with it. They want to go shoot an ILS and wonder why the instructor is dragging his or her feet.

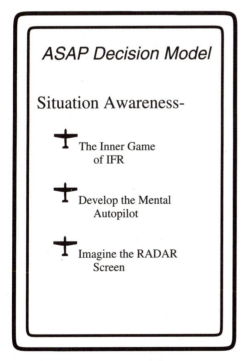

Fig. 5-2 *Situation awareness component of ASAP decision model.*

A good instructor, however, will make it clear that this basic attitude instrument practice is the stepping stone to that ILS approach and the real world of IFR flying.

Much has been said about how a pilot should view the instruments to build a mental image of the airplane's attitude and position. Certainly some flight instruments should be viewed more frequently than others, but I cannot say that there is a single path your eyes should take around the panel that will always present the best picture. The instrument scan or instrument cross-check is a personal thing. I cannot tell you how I do it, and even if I could, it might not be right for you. And there could never be a standard way that would work in every flight situation. Straight and level will require emphasis to different instruments than a descending, turning, radial intercept will. The instrument scan that you use can only be developed with practice—lots of practice. So do not get bored with basic attitude instrument flying; use it to create your technique.

Your technique will become extremely important because eventually everything else will rely on it. In the early going it takes just about all your mental energy to hold a heading and altitude while under the hood. You will often get caught doing the "fighting fires method." You will be concentrating so hard on holding your heading that the altitude gets away. It's like a small fire has ignited at the altimeter, so you rush your attention over to the altitude to put out this fire and once altitude is regained you notice that another fire has erupted back on the heading. You go back and forth, never fully controlling either one. While all this is going on, you are totally oblivious to everything else that is going on. You really do not know where you are. You do not have time to prepare for an upcoming approach. You cannot find the proper frequencies because the airplane gets away each time you look through the chart book. The controller is talking to someone, but it does not register that he is talking to you. Under these conditions, you cannot plan ahead. It is all you can do to keep up with the present; it's just too much to ask for you to be looking to the immediate future. You are mentally saturated, and you will make silly mistakes one after another. When it's all over you will say, "I knew better than that," but under the stress of the moment you did not have the time to use your best judgment.

To be effective and safe, pilots must move away from this ball of confusion to mental control. As the instrument scan improves, it will take less and less mental energy to fly the airplane. This leaves more and more time to think ahead and plan. It is like walking and talking with a friend. Your main attention is on the conversation. You are listening and responding. You are giving deliberate thought to what you are saying, but while this is going on you are walking. Occasionally you come to an incline or a step to climb or need to make a turn to avoid a tree. Your brain is taking in this information, and your legs and feet are adjusting their steps, and making the turns without stealing much mental capacity from the conversation. The brain is handling the walking and the talking simultaneously. The walking part is almost instinctive and not using much brain power. This leaves the brain the room to analyze what is being said, and you become a brilliant conversationalist. Pilots must get to the point where their flying of the airplane is almost instinctive, and the physical demands of flying the airplane do not require much brain power. This leaves the brain more room to put money in the bank, ask what-if, and stay ahead of the airplane. The airplane goes on a "mental automatic pilot."

This is the key to everything else because you cannot do anything else if the airplane itself takes up all the space on your disk. Past this point a whole new world exists for the pilot. It is a thinking level not a physical level, and it is here where expert pilots work.

Imagine the radar screen

The first time I sat and watched an air traffic radar screen, it was like opening up a whole new eye. For the first time I could see the big picture and understand why controllers did what they did. Every pilot should take advantage of operation raincheck, which is an FAA opportunity for pilots to visit their local control tower and radar room. It is hard to ask a pilot to imagine what the radar screen looks like if he or she has never seen one in the first place.

When I am in flight, I picture myself as moving across the sectional chart or on the radar screen. Then I try to build the most complete picture. I place myself on the screen in the proper position relative to the approach I'm preparing to shoot or the airport I am inbound for. Am I south of the localizer? Am I outside the compass locator? When I know these things, I know what direction to expect turns from ATC, and I will know instantly when a controller issues a turn that will not work or is incorrect. I will know when the controller is late with the call. Then I try to place other traffic into my mental radar screen. I hear the controller tell another pilot to fly 360 to make an intercept.

From this I know which direction it is coming from. I hear the pilot say "King Air 24P" or "United 317," and I instantly know that these airplanes are faster than my Cessna 172. An instrument approach is like a big funnel. Airplanes are vectored to the entrance to the funnel, and then they fly through the funnel to the airport using the approach procedure. Imagining the radar screen, I can get a feel for how many other airplanes are being positioned for the entrance to the funnel. This technique is valuable for both IFR and VFR flight. I can imagine airplanes in motion around a traffic pattern as well. If another airplane reports on the downwind leg of the pattern while I am turning crosswind, I know that when I am on downwind, in just a few seconds, that I should look for the other airplane turning base leg. This keeps us separated, and the traffic moves smoothly.

I was flying into a fairly busy Class D airspace recently. This particular airport has an ILS approach and an 8000-foot runway. The airport also has a National Guard helicopter unit, a regional airline base, a large air taxi operation, and a steady flow of cargo traffic. All this makes for an interesting mix of traffic. I was flying in VFR conditions with a pilot who was working toward her flight instructor certificate, and we were making the flight specifically to practice imagining the air traffic. We started listening early to the control tower frequency so we could build the big picture. Before long we were threading our way between someone on that ILS approach, another inbound from the south, and another who was already in the traffic pattern. The Class D controller, who had no radar, advised us of the traffic lining up for the ILS. My student asked if I saw the ILS traffic. Without turning my head I pointed to the right and said he is right over there. The student looked in that direction, did not see the other airplane, and then said, "Where do you see them?" I said, "Oh I don't see them with my eyes—but they are just about there," pointing again. I never actually saw the airplane, but I knew that it had not been given its final approach turn, so we must be inside or closer to the airport than it was. Because I could "see" this, I knew that if we kept our speed up, we would arrive at the runway before that approach traffic and that we should just keep right on coming. If we had delayed any, a conflict would have been created. The mental radar screen helped build and keep an image of what was taking place, and that kept us ahead of our airplane and also the flow of traffic.

There are countless other examples. As you fly you can learn more from listening to the radio than you can talking on the radio. When inbound to an uncontrolled airport in VFR conditions, plan to switch to its AWOS frequency first and then its advisory frequency at least 20 miles out. When you first get on the advisory frequency, do not say anything; just listen. In a moment you may hear traffic at your destination airport making traffic reports. From this you will know the active runway. Knowing the active runway, you can maneuver to enter the traffic pattern with the smooth flow of traffic. I am in the traffic pattern all the time and usually with two and sometimes three other airplanes. This means that 30 seconds will never go by without someone making a position report: "Jonesville Traffic, N1234A left downwind runway 27," "Jonesville Traffic, N4321Z left base for 27," and so forth. It should be no mystery to anyone approaching, who is listening on the frequency, which runway is in use and that there are many airplanes in the pattern. But still you hear inbound pilots say, "Jonesville Unicom, this is N6789B coming in from the south, requesting your active runway and traffic, over." Where has this guy been? And what is worse, the guy on the ground who happened to be passing by the Unicom radio and who knows nothing of the traffic in the pattern will answer the inbound pilot (usually slowly) "N6789B winds are favoring runway 27 and we have had some airplanes in the pattern today." All this is taking place while airplanes are in motion in the pattern. This conversation is blocking the frequency and does not deliver information as valuable as if the inbound pilot had simply kept quiet and listened for the information.

Staying ahead of the airplane is easy to say but hard to do. One of the ways it can be accomplished, in any flight environment, is to use your imagination. Start to see the radar screen, or traffic pattern, in your mind's eye.

Play the mental game

The major concept of Timothy Gallway's *The Inner Game of Tennis* is that tennis players reach a high level of skill when the winner has to rely more on thinking skills than on physical ones. When I play tennis, I usually beat myself. I hit the ball out of bounds or into the net, or I never hit it at all. But when two really good tennis players get together, neither can rely on the other to give away the game with poor shots. At the level that those players are on they can hit the backhand, forehand, lob, or any other shot you can think of. So how is it possible to beat a player at that level? The only way is to outsmart the other player. You must think better than your opponent. You must have a plan, a strategy. You are thinking two and three shots ahead. If you and I were to watch two players on that level, we would be seeing two physical athletes in action, but to the athletes themselves see two minds in action. At that level the action is beyond the physical and is controlled by mental skills.

Pilots and controllers play a similar mental game, but like really good tennis players, it is on a high level. Many pilots do not know that this mental game even exists, and they miss the hidden messages. I heard a controller ask a single-engine VFR pilot inbound to a busy airport what his approach speed would be on final. The controller was a veteran; this was not his first day on the job, and he already knew approximately how fast a single-engine airplane would fly on final, so why was he asking? He was not really asking a question; in reality, he was sending a message. He was telling the single-engine pilot that he would have to fly as fast as possible if he had any hope of blending in with the faster traffic. The question was actually a code, a subtle hint to keep the speed up. But the single-engine pilot did not get the hidden message. He said, "Oh, about 70 knots." Because the pilot was not playing the mental game, he was penalized with long, out-of-the-way vectors, and the pilot never knew why.

Sometimes controllers will take advantage of pilots who are not in the game. If there is ever a conflict that will require one airplane to be delayed, the controller will pick out the pilot that he or she thinks will not know any better or the pilot who can be intimidated into not questioning the call. That pilot will be assigned a heading that seems to go nowhere, a delaying tactic. Have you ever been given a heading that was not the direction you wanted to be going and it seemed the controller simply forgot about you? Of course it is possible that the controller just forgot, but most likely he or she sent you out of the way and was just hoping that you were too timid to ask what was going on. When controllers make this deceptive heading assignment, they will "sell the call" with a clear, and assertive voice. The tone sounds so emphatic that it must be important, so important that you do this even if it makes no sense. It is much like when a baseball umpire will sell the call. When there is a play at the plate, everyone wants to know if the runner is safe or out. Whether or not the call is correct the umpire will yell out the call and make a big production with his hand signals because he knows if he looks like he is sure, he will get fewer boos and manager arguments. Controllers who sound like they are sure will get fewer pilot questions, and this buys them time until the conflict has been solved. The next time you are put on one of these limbo headings, politely ask "Approach, about how long will you need us on this heading?" This will be a subtle message back to the controller that you are indeed playing the game and he or she should pick on someone else who is not. The controller will never let on that he or she has been found out and will say, "You can expect a turn on course in 5 miles," or "I have to coordinate that with center," or something else that makes it sound like there was a reason all along. Of course, there are times when controllers have no choice but to place airplanes on headings that are not to the pilot's liking. Do not get the idea that this mental game is adversarial, but when the controller knows that you are playing the game, he or she will give you some respect and tell you what the problems are.

Controllers make an evaluation of a pilot within the first few radio transmissions. They can tell pretty quickly who they are dealing with. Conversely, pilots playing the game will know when the controller is a little unsure and not in the game. When you are flying and playing the mental game and you have a controller who is doing the same, the result is a completely professional atmosphere. You can sense when the controller's response is urgent and you make your own responses quickly, to the point, and urgent. You can anticipate what the controller's next move will be, and you are all over it when the request comes in. Playing the game, you can see the conflict that the controller is confronted with and you can work together to solve the problem. How many times have you heard a controller say to a pilot "thanks for the help" or a pilot say "thanks for the work"? That is a recognition that two pros were playing the game.

Playing the mental game may be situation awareness at its highest form. Knowing what to expect next helps the pilot get prepared for what will happen next. Staying prepared keeps the pilot in a position to make quality decisions.

Action

Many pilots are not confident of the decisions that they make. Nobody wants to make a mistake, but some pilots are so afraid that their decision will be wrong that they do nothing at all. There are many situations where any decision would be better than no decision. Being unable to take action on decisions (Fig. 5.3) is the same as not being able to take control as pilot in command.

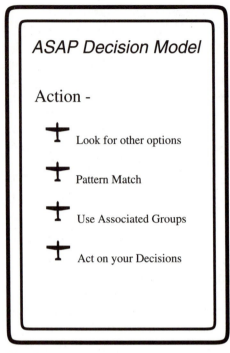

Fig. 5-3 *Action component of ASAP decision model.*

I had a flight instructor applicant on his checkride once who failed the test because of his inability to make a decision or to carry out his decision. The oral examination started at 8:00 a.m. and continued until 11:30 a.m. This applicant had done extremely well. At 1:00 p.m. the applicant and examiner met at the airplane and the flight had gone very well. It was only at about 3:00 p.m. when the test was all but over that the problem came up. The two were flying west back to the Raleigh-Durham airport. The approach controllers told them to make a straight-in for runway 32, but there was a patchy area of clouds that blocked the view of the airport. My applicant could not see the runway and therefore did not start the descent to traffic pattern altitude. Soon the airport came into view, but now they were too high to make the straight-in approach. The applicant had flown for approximately 5 minutes without deciding on a course of action to solve the problem, and then it was too late. The examiner ultimately recommended a course of action. The applicant could have told the controllers about the clouds. He could have asked for a vector around to a position where he could see the airport, and if in fact the airport had gone IFR, he should have been planning for an instrument approach. But he did not do any of these things. Momentarily he stopped being in command. The examiner correctly assessed the situation and ruled that the applicant had not controlled the situation to the degree required of a flight instructor. If a student had been on that flight instead of an examiner, the student would not have stepped in and saved the day. It was a tough lesson learned about being and remaining in control. When I talked with the applicant about it later, he said that he had considered several possible actions, but he was unsure about what was best and time slipped by while he debated it. The examiner said later that just about anything that he might have done would have been better than what he did do—which was nothing.

All pilots must develop within themselves a level of confidence about their flying that will allow them to act and not be paralyzed and prevented from acting. When you have no confidence in your decisions, you become unsure and afraid to act on them. You cannot be so afraid of doing something wrong that you cannot do something right. Some pilots are reluctant to act on their decisions because action requires the pilot to be bold and assertive. They worry that if they are bold but mistaken, they will be embarrassed.

A young first officer at a regional air carrier was calling out the airplane's altitudes down an ILS approach to his captain. "100 feet to decision height…50 feet…20 feet above DH…decision height…Captain, we are at decision height with no runway in sight…Go around Captain…My airplane! I'm going missed approach!" The first officer reached over and pushed the throttles forward before realizing that he had misread the altimeter and in fact they were still 1000 feet above decision height. The first officer had been bold, but mistaken, and then embarrassed. But wasn't it better that the first officer was ready to take action? What if the airplane had actually been at the decision height and the first officer failed to act by doing nothing until the airplane hit the approach lights? I believe that there will be more circumstances where it would be better to be assertive and incorrect than passive and inactive. Of course it would be even better to be assertive and correct. Being correct is where all your study, flight lessons, and experience pays off, and being assertive is where your confidence pays off.

If the conflict comes down to assertive action or fear of making a mistake, how can we be less fearful and more assertive? Some keys to tipping the balance toward more confidence and more assertiveness are to use the points of the ASAP model so far and then add these: pattern matching and the use of associated groups.

Pattern match

A pattern match takes place when a pilot facing a decision matches the proper solution to the problem. Pattern matching is very applicable in time-stress situations. Remember that time is a luxury to the pilot. Many tough decisions will present themselves when the workload is greatest and therefore the pilot has the least time to deliberate over the solution. One reason that experienced pilots make better decisions is simply that they have seen the situation before and they can remember what worked last time. They are not necessarily smarter than the inexperienced pilot; they just have been there before. The experienced pilot matches what worked last time to the problem this time and looks like a genius. If inexperienced pilots could pattern match like a pro, they would also be a pro, but how do they get this skill without years of different experiences? It is the age-old conflict. You can't get the job without experience, but you can't get the experience without the job. There is no 100 percent effective substitute for experience, but one technique that inexperienced pilots can use to make decisions like experienced pilots is to pattern match using associated groups.

Use associated groups

An associated group is a family that goes together. In an airplane these families consist of interrelated systems, instruments, and procedures. In most airplanes, for instance, the landing gear and the hydraulic system are in the same family. The outside air temperature gauge and the pitot heat switch are related and therefore part of an associated group. The airspeed indicator and the stall horn go together. The exhaust gas temperature gauge and the setting of the mixture control are associated. You could come up with many more examples. When a pilot faces a dilemma, the first question that must be answered is, Which family does this dilemma affect? This will direct the pilot to inspect the instrument that will best tell about the problem and lead the pilot to the best corrective action. The pilot must make these logical associations and know enough about the elements of these families to understand why and how they go together. If a directional gyro fails, it would be wrong to check the electrical system because most directional gyros are not electrical. So the pilot must have sufficient background knowledge to at least know what goes with what. Past that baseline knowledge, using associated groups can help the pilot diagnose the problem.

Act on your decisions

Finally, give yourself some credit for your hard work, your study, and your practice. You owe yourself some confidence. Use that confidence to be assertive with your decisions. Trust your instincts. Use your training. Pattern match using associated groups and act on your decisions. Decisions are a normal function of any flight, so do not be afraid to make them.

Preparation

The P in ASAP is preparation (Fig. 5-4). Preparation is more than staying ahead of the airplane. Staying ahead of the airplane means that you have a good understanding of the situation at hand, but preparation goes even farther. Preparation implies that you are planning for your own destiny.

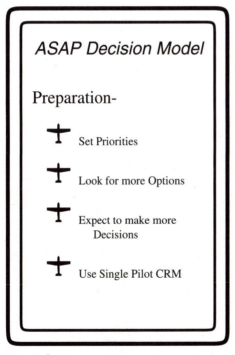

Fig. 5-4 *Preparation component of ASAP decision model.*

Set priorities

There seems to be two pilot workload levels: idle and wide open. Often you can fly for hours without much taking place. These down times can be a trap because your mind can sort of coast when the workload is low. This means that you lose a little sharpness, a little awareness. Pilots have been caught off guard during these times. A problem could slowly be developing, but the pilot feels that everything is all right, and the small problem that is not caught early grows into a monster. I have seen pilots, while en route and at a low workload segment of the flight, just make a sweep of the panel and cockpit. They start on one side and work their way across, looking at each gauge, each switch, each handle, each radio just to verify that everything is where it should be. They make this sweep routinely throughout the low workload portions of the flight just for a reality check and to ensure that no undetected monsters are growing.

The other workload level, of course, is like drinking from a fire hose. You fly in near boredom for an hour and then in the span of 2 minutes a ton of bricks gets dumped on you. During these high-stress encounters, the pilot must prioritize the workload. Pilots must do more than 1 thing at a time, but we cannot do 12 things at a time, so to get all 12 things done, we must line them up in a logical sequence and work them out 1 by 1. As you get closer to your destination, you can count on the workload going up, so get ready. Get the in-range and landing checklists done early. Get an understanding of probable runways in use from AWOS and ATIS. When inbound in IFR conditions, listen for the controller to say the word *expect*. The controller will say "expect ILS runway 2 right" or "NDB runway 18, you can expect that." this is a message to the pilot. The message is not just which runway you will be landing on; the message is actually "Start looking for the approach chart, and start setting up the radios for the approach procedure." This will become a very high priority. If you do not have the proper frequencies set in the proper radios, you will not be able to shoot the approach anyway, so preparing for the approach must be the first item on the priority list. Once that job is done, another item will move to the top of the priority list. Always answer this question: Am I being vectored for a straight-in approach or will I be flying a "full" approach? This is sometimes not stated by the controller but implied in the way he or she handles the traffic. If you are asked to fly directly to the radio fix (VOR, NDB, etc.) that is used to shoot the approach, usually you will need to fly away from the airport first, make a procedure turn, and then turn and descend into the airport. If, on the other hand, you are receiving a series of headings (vectors), you probably are being positioned to fly straight in without a procedure turn. You can expect "full" approaches at smaller airports that have poor or no radar coverage and nonprecision approaches. You can expect faster, straight-in approaches at larger, busier, radar-controlled airports with precision approaches. But if these signals are not clear, you should never hesitate to ask the controller, "Will this be a straight-in or full approach?" Get this understood early and off the priority list.

With the setup to the approach settled, the next priority will probably be briefing yourself on the approach. What are the approach's altitudes? Is this a timed approach? Should I plan on a circling approach? What is the missed-approach procedure?

You can see that if pilots do not attack all these questions in a logical sequence, they will soon all crash together at once and be unable to handle everything simultaneously. The workload would then become unmanageable and the pilot's awareness and safety would be drastically reduced. As the workload begins to increase on any flight, ask yourself what must be done now, what needs to be done but is not pressing, and what would be good to get done if there is time. Line up these workload items in priority, deal with, and then dispose of them one by one.

Look for more options

Don't always accept that the course of action that you are currently on is the only option or the best option. Ask yourself, "Is this the best way to handle this situation?" Do not get locked in. It is easy to get in a flow of events and just ride along.

I routinely test pilots on navigation skills. A pilot will plan the first leg of a VFR cross-country, and we will take off in search of the first checkpoint. Ordinarily the pilot will have calculated the compass heading, distance, and time to the checkpoint. Once under way, he or she starts to look for the checkpoint, such as a small town with a railroad track. The pilot wants to find the point so much that he or she becomes convinced that he or she is actually seeing the checkpoint. The pilot will see what he or she thinks is the railroad track and strike off in an illogical direction, completely abandoning earlier calculations. The pilot believes something that is not true and starts making decisions based on a false assumption. Soon that pilot will claim to have found the checkpoint, even though the railroad track turned out to be a power line and that line runs east and west through the town when the railroad track was supposed to run north and south. During the time that this was taking place, the pilot was convinced it was the correct thing to do. He or she never considered another option or another possibility. The moral of the story is never to completely trust the option that your flight is currently pursuing. Even when everything is going well, ask yourself if there might be a different way to do this.

Expect to make more decisions

The more I fly with other pilots on tests, the more I get the feeling that pilots are not accustomed to making decisions. They assume that everything will turn out and that only a really abnormal situation would require them to take control with their decisions. To really be prepared to continue the flight, a pilot must change this attitude. Pilots must accept the fact that decision making is normal. If you are going to fly, you are going to be placing yourself in a decision-maker's role. On any flight you expect to take off, you expect to navigate, you expect to land. It should not be a foreign idea to expect to make decisions and to make them one after the other, with each affecting the other. As I approach my destination, I know that I will be facing yet unseen circumstances that will require me to make judgment calls. I need to be in a readiness frame of mind, expecting and eager to take on the inevitable decisions that will be necessary. Being in the decision-maker's position should never surprise a pilot.

Use single-pilot CRM

For many years now crew resource management (CRM) has been a dominant issue with airline crews. In its classic definition, CRM involves pilot teams, who work together to form a synergy. General aviation pilots, who more often fly as the only pilot in the plane, have always felt a little left out of the CRM craze. But being prepared means having all the facts surrounding a situation, and single pilots can use CRM techniques to gain these facts. We should rename CRM, *cockpit* resource management when only one pilot is involved.

When preparing to make an important decision, the conventional wisdom is to do some research so that the decision can be made from a position of understanding. When faced with deciding where to attend college, people usually travel to a school's campus. They talk to the teachers, look over the facilities, and read the course offerings and about majors. They want to get an understanding of all the factors that will affect the decision. They know that a better decision will result when they have all the outside facts.

Pilots facing an important decision will also achieve a better result when they decide based on knowing all the facts. CRM for the single pilot used to mean organizing the charts and checklists in a way that would be easy to reach while in flight. Today single-pilot CRM means using your radio to acquire knowledge so that better decisions will result. As long as I have my radio, I have a large crew. I can call flight watch or a flight service station to get a word picture of weather radar. I can switch frequencies to an upcoming airport's AWOS to get a current report, and I can communicate with controllers or other pilots to work out the sequence of traffic. One time I could not get the landing gear down through the normal means, so I called the Unicom frequency and soon had an airframe and powerplant (A&P) technician on the radio and in my crew. In fact, I often have more people on my crew than would fit in the airplane. I have asked a controller to give me a localizer frequency. Yes, it was on my chart, but at the time I was bouncing through dark, wet, and turbulent air and knew he could give it to me faster than I could look for it. If another pilot had been onboard, I would not have hesitated to ask the other pilot for the frequency, so in a way I made that controller my copilot. I will ask any question of anybody to build my understanding of the situation, because I do not want to make a poor decision and later say, "I really wish I had known that before."

By placing people on the ground in your flight crew, you can be better prepared to make decisions. Considering all the factors that surround a decision will definitely yield better results than working alone and "shooting in the dark."

6

Pilot Categories

When you want to solve a problem, it helps to know exactly what problem you wish to solve. No one remedy will help every pilot. Flight instruction has always been individualized and one on one. A good flight instructor will work to solve an individual's problems and weaknesses with a tailor-made remedy that is designed for that one flight student. Likewise, I wanted to be able to identify problems within pilot groups so that specific training strategies could be used to solve specific problems.

Over time I have arrived at a set of pilot categories that are representative of pilot behaviors and decision-making tendencies. These categories came from years of pilot observation and from the research detailed in the Appendix. These observations are of fully certificated pilots. None are from student pilots.

After grouping them, I named each category based on a phrase that describes the group. The categories became

- Information managers
- Nonassertive decision makers
- Snowball effect
- Lost in space

There are also two subgroups: illogical decision makers and good decision makers/poor fliers.

As you read through the characteristics that make up each category, you certainly will see yourself on occasion. If you are a pilot, consider this a self-examination. If you are a flight instructor, you can use these characteristics to identify problems with your own students and design strategies to attack those problems.

The information managers

Piloting an airplane was once considered a physical task. It involved the moving of levers, switches, and flight controls. Today the pilot must do much more. The pilot is an organizer, a planner, and a systems manager. The job of piloting has shifted from physical to mental tasks. Pilots who have been able to make this shift are the pilots who fly more safely and with more confidence. Pilots who fly mechanically, simply following instructions and "driving" the airplane around like they would a truck, are dangerous. It is clear that the job of safe piloting is no longer just operating a machine; it is managing information.

I call this group the information managers because it is their skillful handling of incoming information that make their flights safe and relatively uneventful. Members of this group are characterized by their ability to anticipate. These pilots are able to control the airplane/simulator without coming near a mental saturation point. This leaves mental capacity available to think ahead and plan for upcoming events. These pilots never seem to be in a hurry, yet they are always doing something. They never let a free moment go without planning something or doing something that will help them out later. These pilots do all the "extras" and little things that make the job easier. In any flight procedures there are several task layers. There are tasks that absolutely have to be done if the flight procedure is even possible. Then there are tasks on a slightly higher level that, although not absolutely required, make the procedure run smoothly. The third task layer involves situation-awareness management and turns out to be the definition of the information manager category. The information manager is constantly and predictably completing these extra third-level tasks.

The definition of information managers sounds much like the earlier description of expert pilots. This is no coincidence. Expert pilots are information managers and vice versa. Taken a step further, a pilot who is an information manager is also a living definition of pilot in command. Men and women who fly as information managers are role models of piloting expertise. They are what we all should strive to be and what all flight instructors should train their students to become.

Nonassertive decision makers

Pilots that make up this group are good, solid pilots. They have had adequate flight training in the past. They know the rules and procedures, but they are not confident. They are like a sports team playing not to lose instead of playing to win.

Specifically, members of this group are characterized by their inability to arrive at a timely decision and/or trust their decision to be sound. They are sometimes timid and unable to take control of the situation. Many are so unsure of themselves that even when decisions are made, they have no confidence in their decision and often change from their first course of action several times. Most of these pilots fly the airplane well and do not seem to be saturated with the physical tasks of operating the airplane controls. They have the mental time necessary to make a decision, yet they have little or no confidence to carry the decision out. As a result they cannot form long-term plans to get out of trouble.

Often pilots who are nonassertive decision makers will attempt to exit the role of pilot in command. They will solicit instruction, tips, or hints from air traffic controllers, flight instructors, other pilots who are riding along, or even passengers. Often their communications with controllers is placed in the form of a question, as if they were seeking a confirmation that the decision they are considering is plausible. This always creates an uneasy conflict with the controller because it is always the pilot who is the decision maker. One of the scariest questions that can be posed to a nonassertive decision maker from an air traffic controller is "What are your intentions?" Many times these pilots will not have anticipated that a decision is eminent and therefore have no answer. The question "What are your intentions?" is the controller language for "What is your decision?" Most decisions made in flight do not come with much time to deliberate. When the controller asks, "What are your intentions?", the controller is actually saying, "Your time to make a decision has now run out—what do you plan to do now?" This lack of decision anticipation can place pilots under a great deal of pressure, and sometimes hesitation of speech, slurred words, and illogical actions follow (see the section on illogical decisions).

There is also a negative carryover from their instrument training. When asked "What are your intentions?" in a critical situation, these pilots act surprised that the controller is asking them to make a decision. They act as if their past instructors had made all the decisions for them during training flights. They do not fully comprehend what is entailed in the phrase *pilot in command*. Even though their safety is at risk, they do not want to assume responsibility for decisions.

They cannot or will not take control of the situation for fear that any decision they might make will be the wrong decision.

It should be understood that the roles of air traffic controller and pilot in flight are well defined. The pilot's title in this circumstance is pilot in command. The Federal Aviation Regulations are clear that the final decision in any circumstance is with the pilot. When a controller "assigns" an instruction, course, altitude, or route to a pilot, it is left to the pilot to accept or reject that assignment. Pilots and controllers do work together, but a controller cannot tell a pilot to fly to a particular airport or even to fly a particular course to an airport. The reason that the power is with the pilot is logical. It is the pilot whose life is ultimately at stake, and therefore it is the pilot who is ultimately responsible.

Nonassertive decision makers do not always respect this pilot/ controller relationship. Often they expect the controller to provide guidance and in doing so abdicate their ultimate decision authority. Air traffic controllers will make it clear they are asking questions. They do not accept responsibility for the pilot's decisions and make it clear that they are there to help, but not to fly the airplane. Nonassertive decision makers tend to talk during flight more than information managers do, but most of the conversation has the ultimate goal of either soliciting suggestions from the controller (or instructor) or confirming a decision they are unsure of.

The snowball effect

This group of pilots is characterized by being "behind the airplane." These pilots are aware of what is going on but cannot keep up with the workload. Very often the reason they do not keep up is directly due to their lack of preparation and wasting of time. These pilots will hear "expect ILS 32 approach" from the controller (which is controller jargon for "get ready!") but will wait up to 5 minutes to select that approach's frequencies and otherwise set the cockpit to be ready for the approach. Snowball pilots do not anticipate. They are reactive rather than proactive. The idea of the snowball is that it starts off small, but as the snowball rolls down hill, it gets larger and larger. The analogy to the pilots of this group would be that a small mistake or oversight causes them to first get slightly behind the demands of the workload, and then they can never get caught up. Unfortunately, these pilots routinely miss a radio frequency change or fail to comprehend a weather report or fail to look at an item on an approach chart, not because they did not think it important, but because they simply did not have time to do it. By the time one item is taken care of, two other items should have already been addressed, and by the time they get around to dealing with those two items, six others will be overdue. Members of this group struggle between the physical demand to control the airplane and the mental demands to think and plan ahead. After a particularly rough flight, many will make comments like "I knew better than to do that," but they were simply workload saturated to do everything or think of everything. Much of these problems are self-inflicted. Because they are unable to take in new, incoming information and utilize this information in a timely manner, they constantly are making the task harder for themselves.

The pilots from this group have this in common: Their flights are often a constant, frustrating struggle for them. It can be like watching a person frantically treading water only to eventually lose the battle. Mistakes are made by these pilots not because they do not know any better, but because they do not have enough time to get to it. These pilots seldom if ever get past that first task level, the tasks that are the absolute minimum. These pilots rarely have time to get to the second-level tasks, and they are so burdened mentally that they never thought of any third-level tasks, what has been called the "extras." These pilots will have normal, stress-free flights from time to time, but the characteristics of the snowball category will reappear when pilot workload increases and weather decreases. Examples of the mistakes I have seen snowballers make are

1 Not properly setting up radio frequencies.
2 Not anticipating a course intercept and flying through the course to the other side, which in turn required a reintercept, and a loss of time that could have been used doing something else.
3 Not aware of position relative to an instrument approach or airports.
4 Not properly setting up headings.
5 Rough control of the airplane.
6 Overcorrecting for courses and altitudes.
7 Failure to descend once established on an approach course.
8 Missed radio calls.
9 Flight past a missed approach point with no action taken.
10 Requesting one particular type of approach but tuning in the frequency for a different approach.
11 Failure to reduce speed and consequently flying the approach faster than the en route speed.

12 Failure to report passing certain points when asked to do so by the controller.

13 Misunderstanding headings, for example, assigned the heading of 020 but flying 200 instead.

14 Not making calculations for time or cloud heights.

15 Misreading the approach chart instructions.

16 Not finding time to even look at the chart.

17 Not finding a particular chart in an approach book. The charts are arranged in alphabetical order, but when mentally saturated with workload, it appears they cannot remember the alphabet.

18 Tracking to a radio station, but upon arrival being unprepared to act beyond the radio station.

Pilots in the snowball effect group do not offer many quotes while the flight is in progress; they simply do not have time. What is said is usually broken sentences that trail off as their mind races to something else. Once a pilot said, "If the glide slope were and I..." followed by silence.

The frustration of knowing what to do but not being able to react fast enough to do it can change some pilots' mood. Some have responded to controller instructions with an angry tone of voice. I saw a pilot become so frustrated when he could not find an approach chart and position it so he could see it that he ripped the book.

Watching snowball pilots can be painful. I always feel like I want to throw them a life ring. The real scary part is that in some cases I know I am watching what could be the last minutes before a fatal accident. I am watching the last minutes of a person's life.

It can also be like watching a person juggling three balls and you throw them a fourth—the problem is that their best is only three. Many of these pilots can either make decisions or fly the airplane, but they cannot do both. There is a direct negative correlation between tasks accomplished and aircraft control. When they encounter a distraction, their airplane control suffers.

The lost in space category

The name of this category should speak for itself. The pilots of this group are characterized by being oblivious to the factors around them. Do not misunderstand this characterization; these people are good "stick and rudder" pilots, in that they can fly the airplane well, but they easily get in over their head. These pilots simply drive the machine (airplane) around with no comprehension of their surroundings. They have little or no situation awareness. Points of decision during a flight may arrive, and they may be unaware of their existence. It is not that these pilots make poor decisions; the problem is that they sometimes do not even know that a decision is called for. They make no correlation between actions that are going on around them and the consequences of those actions. They will get into real trouble and never even know they were in danger.

Other problems that are customary to this group are

1 Improperly switching a navigation radio when they are using that radio to navigate.
2 Repeating instructions back to the controller by rote, but then not carrying out the instruction that they had just repeated.
3 Failure to prepare for an upcoming flight procedure.
4 Losing position awareness on the approach, leading to a failure to descend on the approach course at the proper time.
5 Consulting the wrong approach chart when setting up the radios for an approach.
6 Once a pilot flew the approach at the destination airport without the glide slope, made a missed approach, asked to divert to an alternate that had a full ILS approach. When arriving at the alternate, he elected not to use the ILS approach although it was available and was the only approach that could safely get the airplane below the clouds. He never understood the implication of losing the glide slope at the destination, so he did not think it a problem not to use it at the alternate.

Illogical decision makers (subgroup)

Many pilots, when faced with a decision, will make an irrational or illogical decision. Once a pilot faced with an alternator failure in the clouds requested a holding pattern. One got lost on a cross-country flight. As the sun went down, the pilot decided to land in a field rather than at an airport that was 6 miles away. When under stress, pilots can make decisions that at the time seem perfectly sound but later loom to be illogical. Pilots who can be classified into the illogical decision makers group also display characteristics in common with either the snowball effect group or the nonassertive decision makers. Therefore I consider the illogical decision makers a subgroup because their illogical solution to the problem may have been their lack of assertiveness or workload saturation.

Good decision makers/poor fliers (subgroup)

A small group of pilots just do not fly the airplane very well. They can be rough with the controls or at times it seems like the airplane is in control and that they are just along for the ride. Examples are a failure to descend on an approach that if flown properly would have ended successfully. Another example is a pilot landing in a crosswind without taking crosswind corrective action and landing the airplane with a "side load" or even running off the side of the runway. I have known a handful of pilots who can make sound judgments and good decisions, but their lack of flying precision prevents them from ever using their judgment.

The value of pilot grouping

The grouping of pilots into categories should not be misunderstood. Remember that pilots are all human, and human performance will vary from day to day and even hour to hour. The characteristic descriptions of the pilot categories never represent a single pilot. The descriptions are a composite of many observations. Any pilot taken individually can have performance characteristics from several different groups.

So what is the value of grouping? It is very hard to solve problems and make improvements until you know what the problem is. As a pilot you know many of your own weak areas, and it is hoped that as you read about the different pilot categories, you saw some of your own thoughts and actions played out. This means that you have further defined a problem or weak area. I can remember many of my own flights where I got behind the airplane or was not as aware as I should have been. I always learn more from other pilots than they learn from me. I have been able to see problems develop from a vantage point that few people have ever had, and I know I am a better pilot because of that view. If you are a flight instructor, you might use the categories to identify problems and issues with your students. Please do not "pigeonhole" your students with the categories. Don't ever say, "My 2 o'clock lesson this afternoon is a real snowballer!" Instead, look for the problems that the categories contain and use them as a tool to teach with a better strategy.

We pilots can be our best critics. If you saw yourself in any, or all, of the pilot categories, you saw some weaknesses. Resolve to seek out a caring, qualified flight instructor and be specific about what you wish to work on during your next flight review or instrument proficiency check. Talk to your local safety program counselor about some targeted ground and flight training. If you do not know who your nearest counselor is, call your state's Flight Standards District Office. Every state has a safety program manager (SPM). The SPM wears the "white hat" in the FAA office. SPMs are never involved in enforcement action. They truly can help out with counselor referrals, safety programs, and plenty of materials. But the SPM is just one source. Being pilot in command is much more fun than flying around wishing you were in command. Do not go another year thinking that someday you will improve or get up to date or finally learn a procedure. Get over the hump. Start becoming a pilot in command.

7

Making Improvements

In the previous chapter pilots with similar tendencies were grouped in categories, and it painted a bleak picture. How can a pilot learn from these tendencies and improve? From observing pilots, and teaching pilots, I have seen three very distinct areas where improvements are needed. Some of these areas are unfortunately built into our current system of flight training. Others are problems based on myths, and still others come from a lack of basic pilot knowledge. The three areas are (1) negative flight training carryovers, (2) declaring emergencies, and (3) pattern mismatches. Problems in all three of these areas have caused accidents, but the good news is that improvements can be made once we understand the problems.

Negative flight training carryovers

Flight instructors have a very tough job. On the one hand we must teach students the very basics in a closed environment. We must get away from other traffic to practice. We fly away from the airport out to the practice area where we learn the airplane's tricks. But then the flight instructor must also teach to the real world. We must eventually get in with the other traffic and deal with real-world problems. Working with pilots, I have seen constant signs of what I call "negative training crossover." Crossovers, as I see them, are instances where a pilot's previous training crossed over into a real-world scenario. Often these were positive and quite helpful to the pilot. A pilot, when faced with an inoperative radio, often remembered to try another radio. But there is a disturbing number of negative crossovers. What the pilot remembers from the original flight training, or the original training environment, can be dangerously out of place in the real environment. The difference between a pilot who relies on the training environment and a pilot who is comfortable with the real environment seems to be *seasoning*. Expert pilots, for example, are savvy.

They know the ropes, the short cuts, the real workings of the system. Nonexpert pilots show signs of this seasoning at times, but many are like first-graders tossed into a class change in high school. They are no match for the real world.

I think this is an area of flight instruction that needs big improvement. I am not talking about the practice of only teaching to the flight test, although that does happen. Pilots who are paying the bill and instructors who want to move forward quickly will often gravitate to teaching just what they know the local examiner is likely to ask. The end result is a pilot who meets minimum standards and by definition then has minimum skills. But what I see goes beyond even this. I see a fundamental problem with the way we teach the topics.

The problem exists within all pilot training, but for now let me target instrument pilots. Here are some carryover problems that exist with IFR pilots in training today.

When pilots train to become instrument rated, the flight instructor often must simulate instrument weather conditions. The flight instructor has the student wear a "view-limiting device" (Fig. 7-1). The objective of this device is to prevent the student from seeing out the window of the airplane. This forces the pilot to view only the flight instruments and therefore simulates flying in the clouds. The device is a hood worn over the student's head or a pair of glasses that are frosted over except for a small clear spot at the bottom of the lens. When flying in the clouds, the pilot has no outside visual reference and must rely on the flight instruments to determine the aircraft attitude and position. The view-limiting device accomplishes this even though the aircraft is not flying in the clouds. The student is therefore given an instrument flying simulation during noninstrument weather. There exists a problem with the simulation, however. The student's simulation needs are met with the view-limiting device, but the air traffic controllers are still functioning as if the airplane is flying under noninstrument rules. The way in which the controller will deal with airplanes is different when actual instrument conditions exist. This forces an additional need for flight instructors to simulate not only instrument conditions but also instrument circumstances. It takes a skilled instructor to do this, and even with quality instruction it may not be clear to the student in training what part of the flight lesson is like a real instrument flight and what part is simulated to accommodate noninstrument weather.

Fig. 7-1 *A student wearing a view-limiting device.*

When flying to a destination airport in the clouds (real instrument conditions), a pilot cannot know in advance if the approach will be successful. The lowest approach altitude may be under the clouds or it may still be in the clouds. The pilot will not know for sure until the approach is flown. Most pilots in general aviation do not have "autoland" equipment, so the airplane cannot be landed unless the pilot can see the runway. The pilot can only see the runway if he or she can descend below the clouds. At the beginning of the descent it is not known whether or not the clouds will be penetrated, and therefore, the pilot cannot know in advance if a landing is possible. In this situation the controller will "clear" the airplane to make the descent and approach to the runway and all hope for the best. The pilot knows that he or she needs to be equally prepared for a missed approach if the runway remains hidden by the clouds or a landing if the airplane emerges from the clouds. The pilot must anticipate the possibility of the missed approach while hoping a missed approach will not be necessary. The pilot must also be ready to make the decision to miss the approach without hesitation should it become necessary.

When flying in clear air (noninstrument conditions), the flight instructor may want to have the student remain in the view-limiting device as they descend to simulate a low cloud layer. Meanwhile, the controller assumes that the pilot can see the runway and often asks the airplane, "Will this be a low approach or a full-stop landing?" By asking this question the controller is planning ahead. If the pilot makes the missed approach, the controller needs to have the airspace ahead clear of other airplanes to ensure traffic separation. But the controller would not have asked that same question to an airplane in actual clouds. The pilot flying in actual clouds could not, at that point, answer the question anyway. The pilot will do everything possible to land but cannot yet know if that is possible. If the flight instructor, flying in noninstrument conditions, does answer the question, "Missed approach or landing?", as requested, he or she gives away the "point of decision" that the student should have to make. When the student hears the instructor say that they intend to make a missed approach, for practice, the student stops thinking about a possible landing, and it becomes an unreal game. When the controller asks the question, it ruins the scenario.

Furthermore, the student is not now placed in a decision situation. The decision has already been made by the flight instructor, forced by the needs of the controller. This is a common occurrence in instrument flight training. But unfortunately, this situation teaches instrument students not to make decisions. It does not teach students to realize that a decision would have been necessary had the situation not been a simulation. This becomes a real problem as pilots fresh from the training environment are faced with real-world situations.

Many pilots when faced with the missed-approach or land decision simply freeze and make no decision at all. In their flight training they had never been taught to make that decision because the demand of the in-flight simulations always had made that decision for them. When pilots come to this decision point, many become uneasy. I have actually seen them move uncomfortably in their seats, and their speech becomes halted and shaken, and their airplane control becomes erratic. It is always apparent that the inability to decide has gripped them.

When this happens, it is evident that these pilots are from a training environment where decisions were most always made for them.

When they have to make the decision themselves, they are caught completely off guard.

I have heard these radio transmissions: One pilot during a missed approach said, "What are my instructions?" Another said, "Nashville, I've missed approach at Smyrna awaiting further instructions." Still another pilot asked, "Do you want me to make another approach at Smyrna?" And another said, "Approach [control] I can try again or come to Nashville, what do you think?"

These questions sounded more like what would have been asked of a flight instructor rather than an air traffic controller. Of course, the air traffic controller would never actually answer any of these questions. The pilots were hoping the controllers would take command at the same time that the controllers were hoping the pilot would start being in command. It was a hot potato that nobody wanted to claim.

It was obvious that these pilots had never practiced making these decisions. The training environment in which they grew up sheltered them too much. Think about it. There are decisions that must be made every day in the real world that are seldom if ever practiced in the training world. In fact, we actually train students not to decide. No wonder these example pilots could not or would not decide. Many had never had to before.

Practice instrument approaches

Like it or not, air traffic controllers are partners in pilot training. Unfortunately, how controllers handle pilots in training has evolved or, more accurately, eroded over time. Pilots in training, especially in instrument training, are treated differently than other pilots. The Aeronautical Information Manual (AIM) has a section that speaks directly to the problem. Pilots in instrument training will often fly approaches for practice and skill demonstration. The AIM names these operations as "practice instrument approaches." The definition given is, "Practice instrument approaches are considered to be instrument approaches made by either a VFR aircraft not on an IFR flight plan or aircraft on an IFR flight plan." But then the AIM says, "Pilots not on IFR flight plans desiring practice approaches should always state 'practice' when making requests to ATC." This would seem to indicate that if a pilot is on an IFR flight plan that no such statement is required. Why is this distinction so important to the pilot in training? Because the AIM goes on to say, "Pilots should inform the approach control facility or tower of the type of 'practice' approach they desire to make and how they intend to terminate it, i.e., full-stop landing, touch-and-go, or missed or low-approach maneuver." But if the instructor complies with this request, the scenario is blown and the decision is taken away from the pilot in training. One solution would be to file IFR and conduct flight training on an IFR clearance. The AIM only requests pilots "not on an IFR flight plan" to state that they are making a "practice" approach and therefore only requests that non-IFR pilots reveal the conclusion of the scenario.

When an approach clearance is issued to an IFR aircraft, it automatically comes with a missed-approach segment if a missed approach becomes necessary. There should be no need therefore for the controller to ask a pilot on an IFR clearance, "How will this approach terminate?" because a missed approach was automatically issued with the original approach clearance. It is the pilot flying VFR, without an IFR clearance, that does not get an automatic missed approach. The AIM says, "VFR aircraft practicing instrument approaches are not automatically authorized to execute the missed-approach procedure. The authorization must be specifically requested by the pilot and approved by the controller." So, if a missed-approach clearance is not automatic for pilots flying VFR without an IFR clearance, it would be necessary for the pilot to work out in advance how the approach would terminate. But if the pilot is on an IFR clearance, the missed approach is automatic and should be built into the scenario. Controllers who ask, "How will this approach terminate?" to a pilot on an IFR clearance are overstepping their bounds and are therefore not being very good training partners. Filing an IFR and receiving an IFR clearance should go a long way to protect the scenario and take the pilot in training from the training environment and into the real world.

If the controller does ask, "How will this approach terminate?" the pilot or flight instructor should ask for "the option." Asking for the option clearance allows the pilot to make a missed approach or land at his or her discretion. This keeps the decision in the hands of the pilot or instructor. If the instructor does not inform the pilot that they have descended below the clouds, then a missed approach is in order. But, if the instructor tells the student to remove the view-limiting device and, in doing so, simulates a descent through the clouds to a position where the runway is in sight, then a landing can be made. This technique does a better job of teaching pilots to anticipate and execute their decisions.

However, the AIM clearly makes pilots in training a second priority behind "itinerant" aircraft. The AIM states that "The controller will control flights practicing instrument approaches so as to ensure that they do not disrupt the flow of arriving and departing itinerant IFR or VFR aircraft." The answer then is to be itinerant. Do not fly approach after approach at the same location. Nothing could be farther from the real world anyway. After the pilot has learned the basics of how to fly the approach, plan to make short hops where practical. File two IFR flight plans, one out to shoot an approach and another to come back. In this way it will appear to the controller that you are an itinerant airplane inbound, and you will receive real-world handling rather than practice handling.

Flight lesson scripts

It is a common practice in instrument flight training to "script" the lesson. Flying an airplane is expensive, and there is an incentive for the flight instructor to pack as much into a flight lesson as possible. To help save the student some money, a routine flight lesson might involve three or four consecutive instrument approach procedures. This allows the student to practice these maneuvers without much time spent between each approach. This is good economy, but it tends to lead to decision problems. First of all, in real-world operations, a pilot would never shoot the same approach four consecutive times. The goal would be to fly the approach once and land. If a landing were not possible, the pilot could decide to fly to another airport where the clouds are not so low. But in training, this is not done. If four consecutive approaches are planned for the flight lesson, it is clear that all the missed-approach versus land decisions are already made before takeoff. Regardless of weather conditions, the pilot will fly a missed approach on the first three attempts and expect to make a landing on the fourth. Second, as stated before, consecutive instrument approaches can trigger non-real-world handling from the controller. As a result the environment becomes less and less like the real world, and the pilot in training becomes insulated. The pilots are flying nonrealistic missions, and the controller is providing practice, not real handling. After five or six flight lessons featuring this method, the student in training may be able to fly the approach procedure with accuracy but has not been called upon to make any decisions yet.

I always try to deliberately place pilots into the missed-approach versus land decision situation during instrument approaches. This reveals that many pilots are not prepared to make this decision. Pilots in training expect to land on what they think will be the last approach of a flight lesson, and they expect to make a missed approach on all others. The true factors that should be contemplated in making this decision are not present. Then when faced with a real decision, they fall apart. Instead of making a clear decision, many pilots simply do what they have done many times before during their flight training. They will ask to go back and fly the same approach a second time without considering the possibility of an alternate airport. They choose a course of action, not after a careful evaluation of the situation, but from a training habit. It appears that well-meaning flight instructors who had their students' training costs in mind and controllers protecting airspace are inadvertently training students who are unable to make a timely decision, and those who do make a decision do so from habits learned during simulation. Unfortunately, the consequences of this habit in the real-world situation lead them deeper into danger.

The flight training that these pilots relied on let them down. Put another way, the flight training delivered as it was only carried them so far, but not past the training environment. The instruction did not include seasoning.

But it's not all the flight instructor's fault. Students also will subconsciously hide in the comfort of the training environment. I conducted a flight review for a private pilot once, and I suggested that we fly into and back out of the Raleigh-Durham class C airspace as part of the review. The pilot was against this. He argued that it was not really necessary, it was too far, and it would be more expensive. I had only suggested the trip to RDU in the first place because the pilot told me that he was uncomfortable talking to ATC on the radio. Nevertheless he was not going to RDU. Just for conversation I asked what flying plans he had for after the completion of the review. "Oh, I'm flying my family to Disney World in Orlando." From where he lived that would have meant a flight through both Jacksonville and Orlando airspace. This would have placed him in the middle of some of the busiest and most complicated airspace and ATC requirements in the United States (not to mention near the Kennedy Space Center). This pilot was content to stay in the training environment and was just kidding himself that he would be ready for the real environment when he got to JAX and ORL. I convinced him that he really needed RDU. Half the battle is won when pilots accept the fact that their training was sheltered and become open to learning what is past the basics and in the real world.

When students learn in an unreal environment, but perceive that it is real because they have seen nothing else, they are cheated, and the flight environment is less safe. When a student from the unreal or training environment is one day placed in the real world, that student will be unprepared and overwhelmed. The results are dangerous. The unreal world teaches pilots not to be decision makers, and therefore they become unable to be pilots in command. Instructors, controllers, and students need to do better. Look for ways to train in the real world.

Declare emergency

Actual in-flight emergencies are rare, so maybe that is why pilots are not very good when it comes to dealing with emergencies. Being pilot in command, however, means being in charge in good and bad times. All pilots should learn the systems of their particular aircraft. It is not enough to check the oil, sump the fuel, and turn the ignition. A pilot in command is in command of the airplane and its systems. Pilots do not also need to be A&P technicians, but we must be excellent troubleshooters. I have actually heard pilots say, "Why do I have to know this, isn't that a mechanic's job?" What they do not realize is that if something should go wrong in flight, an A&P technician will never see the aircraft unless a pilot figures out how to get safely to the ground. This means that the ultimate job of the pilot in command is to return the aircraft and its passengers safely to the ground even in the face of an emergency.

Emergencies require pilots to utilize all their skills and resources. One resource that the pilot has is the emergency declaration. In the air traffic control protocol, a pilot who has declared an emergency has priority over all other aircraft. In some cases, when it is imperative to get an airplane on the ground, a declaration of emergency is the best solution to the problem. But there is another problem. Almost universally pilots are reluctant to tell controllers when they have an emergency. The more I talk with pilots the more I realize that this reluctance and even fear of declaring an emergency is widespread. In every case the pilot showed reluctance, with the reason being repercussions from the FAA. Pilots think they will lose their pilot certificate or get a monetary fine if they declare an emergency. Pilots freely volunteer their distrust of the FAA. Pilots know when they are not as proficient or not as sharp on the regulations as they should be. They think that if they are not happy with their performance, surely the FAA will not be either. Simply stated, pilots do not declare emergencies for fear that they will be exposing their faults, and they do not trust the FAA to handle this knowledge fairly.

So pronounced and so universal among pilots is this attitude that I contacted the air traffic control support manager at the Nashville, Tennessee, air traffic control tower. This manager and I have known each other a very long time and he has always been helpful. He is also a pilot and can always see things from both sides. From his account, there is a great deal of judgment involved in how controllers handle emergencies. Believe it or not, there really is no hard and fast guideline about what to do when a pilot declares an emergency. In most cases, if the emergency ends with a safe landing, nothing at all happens. The controller will record that a pilot declared an emergency on FAA form 7230-4, which is their daily record of facility operations. There are only three times when the controllers are *required* to notify the local Flight Standards District Office (FSDO). Paragraph 4-1-4 of the ATC handbook says:

> *Notify the Washington Headquarters, and the appropriate FSDO through the Regional Operations Center whenever:*
>
> **1** *The aircraft [that declared an emergency] is an air carrier, a commuter, or an air taxi; or*
>
> **2** *The aircraft is carrying a member of Congress or prominent persons; or*
>
> **3** *The emergency is or may become newsworthy by coming to the attention of the public or the news media.*

When you fly a general aviation airplane that is not an air taxi, if you declare an emergency, the controller is not even required to tell the FSDO. Remember, the FSDO is the certification and enforcement branch of the FAA.

There are also different levels of emergency preparedness. If a pilot reports in that he or she has a rough-running engine and needs to come straight to the airport, the controller will give the pilot a vector and expedite arrival. The word *emergency* is never uttered. The controller may ask if there is any further assistance the pilot needs. The pilot says, "No, I think I will make it to the airport." Even though the pilot has said he or she will make it to the airport and all is well, the controller still may alert the crash, fire, and rescue (CFR) personnel that there is a level 1 emergency alert. All this means is that the CFR people get up from the dinner table or push back from their card game and go sit in their trucks. They open the building doors and start the truck's engine. If the airplane lands without further incident, they shut down the truck, close the doors, and get back to dinner. The pilot never knows this happened. If the pilot declares an emergency, the same thing would happen. The controller would ask if the pilot needed further assistance. If the pilot does not declare an emergency, the controller can do it anyway. This rolls the trucks into position just in case they are needed. It does not cost the pilot any money to have the trucks rolled into position.

When the airplane lands safely, the event is recorded on the daily log and that is as far as it goes. If the controller does anything, the first notification would be to file FAA form 7230-6, which is a Flight Assist Report (Fig. 7-2). When this form is filed, a copy is sent to the local FSDO, but this could be a good thing. A controller can actually receive a monetary award for "outstanding" assistance. The most common examples of this are helping lost pilots get found and preventing a gear-up landing. The Flight Assist Report even recognizes that other pilots can provide flight assistance. Paragraph 4-1-5 of the controller's handbook states

> *Pilot Recognition   1. When a pilot aids in providing flight assistance, the Air Traffic Manager shall review the circumstances, and if appropriate, write a letter of recognition. 2. When the pilot assistance is of an outstanding nature, the Air Traffic Manager shall review the circumstances, and if appropriate, prepare a regional level letter of recognition.*

When the Flight Assist Report reaches the FSDO, it may not do anything with it. In the case where a controller helps a lost pilot, the safety program manager at the FSDO office might locate the pilot who was lost or the flight instructor in the case where the lost pilot was a student. The safety program manager would in that case suggest that additional VFR navigation skills be taught. Otherwise the CFI might never know the student got lost (the student probably will not tell) and therefore not realize additional training was needed. Nobody gets "in trouble"; the safety program manager is the person in the FAA

FLIGHT ASSIST REPORT		REGION	FLIGHT ASSIST REPORT NO.

1. FACILITY	2. DATE	3. TIME (GMT)	4. AIRCRAFT IDENTIFICATION	5. NO. PERSONS ON BOARD

6. POINT OF DEPARTURE		7. ORIGINAL DESTINATION	For Items 8-15 place "X" in appropriate box except where designated otherwise.

8. FACILITY TYPE	FSS	TERMINAL	CENTER	9. OCCURRED DURING HOURS OF	DAYLIGHT	DARKNESS	10. INCIDENT REPORT, FAA FORM 7044, FILED 8020-11		YES	NO

11. AIRCRAFT DESCRIPTION	AIRCRAFT CATEGORY			TYPE		NO. OF ENGINES			DESIGNATION (Specify)
	GEN. AVIA.	MILITARY	AIR CARRIER	PISTON	TURBINE	ONE	TWO	THREE OR MORE	

12. FLIGHT PLAN	VFR	IFR	NONE	13. ACTUAL FLIGHT CONDITIONS	VFR	IFR	VFR OTP	UNKNOWN

14. PRIMARY CAUSE	LOST	LOW FUEL	CAUGHT ON TOP	EQUIPMENT MALFUNCTION			OTHER (Specify)
				COMM.	NAV.	MECH.	

15. PRIMARY METHOD OF ASSISTANCE	RADAR	DF	VOR	ADF	OTHER AIRCRAFT	GEOGRAPHICAL FEATURES	SPECIALIST DETECTED AND ADVISED PILOT

16. BRIEF SUMMARY OF INCIDENT

17. ATC SPECIALIST WHO PROVIDED FLIGHT ASSISTANCE SERVICE		
NAME	POSITION WORKED	TITLE AND GRADE

SIGNATURE (Facility Chief)	COPY DISTRIBUTION

FAA Form 7230-6 (11-73)

Fig. 7-2 *Flight Assist Report (FAA form 7230-6).*

office who wears the white hat—he or she does not have enforcement authority. The Nashville air traffic controller tower reports that they file a Flight Assist Report at the rate of one or two reports per year.

I asked the air traffic support manager, "It seems that most emergencies whether actually declared or not are handled 'in house' but some do go farther. How many declared emergencies ultimately result in a pilot violation?"

"It is rare, very rare," he said, "and the violations usually are due to some other problem other than the emergency that comes out in the questions." He told me the story of two emergencies where pilots eventually were violated, but both violations had nothing to do with the fact that an emergency had been declared. One involved a pilot flying a light twin-engine airplane who declared an emergency and in the excitement landed on a taxiway. The airplane was not damaged, and the pilot was unhurt. In the questioning it was discovered, however, that the pilot was only single-engine rated. The other involved a pilot flying a single-engine airplane who had some sort of engine problem, declared an emergency, but landed safely. It was later discovered that this pilot had no medical certificate. "Both those pilots blamed the emergency for their violations, but in truth the emergency did not get them into trouble; they did that on their own," the controller finished.

I found it very interesting that the Flight Assist Report does not have a check box for "emergency declared." The controller could include that fact in box 16 under "Brief summary of incident," but there are no FAA forms specifically designed to handle emergencies alone.

If a pilot "willfully and maliciously" breaks a regulation and an emergency is also part of the incident, the controller can notify the FSDO of a suspicion that a "deviation" from the rules has occurred. This notification is completed on FAA Form 8020-17, the "Preliminary Pilot Deviation Report" (Fig. 7-3).

I then wanted to know what would happen after notification of an emergency reached the FSDO. I interviewed an FAA flight standards inspector about the subject. I must tell you that this inspector and I have worked together for many years as well, and he is practical and brings common sense to the job. For the most part he dispels the bad reputation that the FAA has earned, but he is only one inspector and may or may not represent the views of the FAA as a whole. The following is the transcript of my interview with him.

> *Craig:* Would you outline what happens when a pilot declares an emergency?

> *FAA inspector:* First, the FAA gets a Preliminary Pilot Deviation Report [Form 8020-17] from the air traffic controller. The FAA inspector working the case then fills out an Investigation of Pilot Deviation Report [Form 8020-18].

Then the inspector calls the pilot on the telephone and hears the pilot tell the story. The inspector will ask about the pilot's total flight time, time in the type airplane, and time in the last 90 days—but these questions are for statistical use and are optional.

Craig: These questions are optional?

FAA inspector: They are used for an accident database. The pilot can choose to answer these questions or not.

Craig: What happens next?

FAA inspector: The FAA can then do one of three things. We can take no action, we can take administrative action, or we can initiate enforcement action.

Craig: How do you decide between no action or administrative or enforcement actions?

FAA inspector: No action will ever be taken unless the pilot is somehow otherwise negligent.

Craig: What do you mean by *otherwise negligent?*

FAA inspector: The pilot took off without radios into IFR or did not follow a clearance, etc.

Craig: Sometimes pilots are faced with malfunctioning equipment. These equipment failures could lead to an emergency. How does the FAA look on this situation?

FAA inspector: Enforcement actions can only be against people not airplanes, so if the airplane breaks, no action can be taken against the airplane.

Craig: You are aware of the distrust that pilots have toward the FAA, and unfortunately this distrust makes pilots believe that declaring an emergency is more hazardous than the emergency. What do you think about this problem?

FAA inspector: When in a life-threatening situation "screw the paperwork." Pilots should not let anything the FAA might or might not do later affect their safety decisions. We would much rather have a safe outcome to any flight. We (the FAA) have a whole lot more paperwork to do for dead people than for live people.

Craig: And in those cases where enforcement action is pursued, what can happen to the pilot?

FAA inspector: The greatest fine that can be imposed on a pilot is $1000 per violation. Is your life worth $1000? The
greatest certificate enforcement action that can be taken is
a certificate revocation and that can drop off in 1 year.

So the paper trail this inspector described went like this: The air
traffic controller can record the emergency in the daily log and
nothing else, they can file a Flight Assist Report, or they can file
a Preliminary Pilot Deviation Report. If the air traffic controller files a
Preliminary Pilot Deviation Report, the information goes to the FAA
inspector who is assigned the case. The FAA inspector calls the pilot and asks the pilot to voluntarily answer some questions for the
databank. The inspector fills out an Investigation of Pilot Deviation
Report and later makes the decision to take enforcement action
against the pilot or let it go.

I was really interested in all these forms. Because they are govern-
ment documents, they should be accessible to the general public, so
I asked to get copies. Figure 7-3*a* is a copy of page 1 and Fig. 7-3*b* is
page 2 of the form air traffic controllers use when an emergency is declared and a violation suspected. Figure 7-4*a* is a copy of page 1 and
Fig. 7-4*b* is page 2 of the questionnaire that an FAA inspector uses
when making the decision to seek a penalty for a pilot or not. Figure
7-5 is the box on the last page of the FAA inspector's form that
records the inspector's decision. An EIR stands for Enforcement Investigation Report. If the inspector checks box A, "EIR initiated," the
pilot is in trouble. If the inspector checks box B, "No EIR initiated,"
the pilot is off the hook.

These are the FAA forms that are used to handle emergency declara-
tions. Do you think it odd that nowhere on any of these forms is the
word *emergency* used? There are no check boxes for emergency declared. The last two forms are for pilot *deviations.* Are we to assume
from this that the FAA considers emergencies as deviations? The FAA
inspector's form is an *investigation* report. The words *deviation* and
investigation have connotations that would make any pilot feel uncomfortable. Okay, it is true that emergencies declared rarely result in
enforcement action against the pilot, but these forms seem to suggest
that he or she is guilty until proven innocent. I strongly feel that the
FAA needs one more form—an emergency declared form. When a pilot declares an emergency, it should not be assumed that a deviation
has occurred and an investigation triggered. Today most of the pilots

	Incident Report Number
PRELIMINARY **PILOT DEVIATION REPORT**	**P** ⎵⎵⎵⎵⎵⎵⎵

Complete and distribute according to instructions on page 3. Complete Items 1 to 9 and 27 to 32 for all deviations; if surface deviation, also complete Items 10 to 14; if air deviation, also complete Items 15 to 26. "ID" refers to FAA location identifiers in the latest edition of FAA Handbook 7350.6, "Location Identifiers." Complete the form by hand or typewriter.

1. Date, Time, and Location of Deviation:

A. Date (Coordinated Universal Time—UTC)
⎵⎵ ⎵⎵ ⎵⎵
M M D D Y Y

B. UTC Time
⎵⎵⎵⎵⎵

C. Local Time
⎵⎵⎵⎵⎵

D. Nearest City or Town and State _____

2. Pilot Information: ☐ Unknown

A. Name and Address

Name (first, middle, last)

Address

City   State   ZIP

B. Telephone Number
⎵⎵⎵ – ⎵⎵⎵ – ⎵⎵⎵⎵

C. Pilot Certificate No. (if military, enter "MILITARY")
⎵⎵⎵⎵⎵⎵⎵⎵

3. Deviation First Detected by (check one):

A. ☐ Error Detection Program (EDP)
B. ☐ Radar Observation (excludes EDP)
C. ☐ Visual Observation (tower)
D. ☐ Flight Service Station
E. ☐ Public, Including Pilots
F. ☐ Other, Specify _____

4. Aircraft Information (complete A or B; always complete C):

☐ Unknown

A. Registration Number (N Number)
_____

B. Flight No. or Call Sign (if applicable) _____

C. Make and Model _____

5. Type of Operation at Time of Deviation (check one):

A. ☐ U.S. Air Carrier (14 CFR 121 or 125)
B. ☐ Foreign Air Carrier (14 CFR 129)
C. ☐ Commuter (14 CFR 135)
D. ☐ Air Taxi (14 CFR 135)
E. ☐ General Aviation (14 CFR 91)
F. ☐ Public Use
G. ☐ U.S. Military
H. ☐ Unknown
I. ☐ Other, Specify _____

6. Type of Flight Rules at Time of Deviation (check one):

A. ☐ Instrument Flight Rules (IFR)
B. ☐ Visual Flight Rules (VFR)
C. ☐ Special VFR
D. ☐ Unknown

7. Phase(s) of Flight When Deviation Occurred (check appropriate boxes):

A. ☐ Taxi
B. ☐ Takeoff
C. ☐ Climb
D. ☐ Level Flight or Cruise
E. ☐ Turning or Maneuvering
F. ☐ Descent
G. ☐ Approach
H. ☐ Landing
I. ☐ Unknown
J. ☐ Other, Specify _____

8. Total Number of Aircraft Involved (if more than one, also provide other aircraft information):

A. ☐ One
B. ☐ Two
C. ☐ Three
D. ☐ Four or More
E. ☐ Unknown

	Aircraft N Number	Flight No. (if applicable)	Aircraft Make & Model
F.			
G.			
H.			
I.			

9. Type of Deviation(s): (check appropriate boxes):

A. ☐ Surface (complete Items 10 to 14 and 27 to 32)
B. ☐ Air (complete Items 15 to 32)

10. Type of Control at Surface Deviation Location (check one):

A. ☐ Operating Control Tower
B. ☐ Nonoperating Control Tower
C. ☐ None, Nontowered Public Airport
D. ☐ None, Private Airport
E. ☐ Unknown

11. Airport Location ID (provide 3- or 4- character ID):
⎵⎵⎵⎵

12. Surface Deviation Type(s) (check appropriate boxes):

A. ☐ Takeoff Without Clearance
B. ☐ Takeoff on Wrong Runway or Taxiway
C. ☐ Landed Without Clearance
D. ☐ Landed or Takeoff Below Weather Minimums
E. ☐ Landed on Wrong Runway, Airport, or Taxiway
F. ☐ Entered Taxiway or Runway Without Clearance
G. ☐ Careless or Reckless Aircraft Operation
H. ☐ Did Not Close Flight Plan
I. ☐ Other, Specify _____

13. Was There a Loss of Separation With (check appropriate boxes):

A. ☐ Ground Vehicle
B. ☐ Personnel
C. ☐ Another Aircraft, on Ground
D. ☐ Another Aircraft, in Air
E. ☐ Obstruction
F. ☐ No Loss of Separation
G. ☐ Unknown if Incursion Occurred

14. If There Was a Loss of Separation It Was (check one):

A. ☐ Under 100 Feet
B. ☐ 100-499 Feet
C. ☐ 500-1,000 Feet
D. ☐ Over 1,000 Feet
E. ☐ No Loss of Separation
F. ☐ Separation Unknown

If Surface Deviation Only, Skip to Item 27

15. Location in Traffic Pattern During Deviation (check one):

A. ☐ Entry or Downwind Leg
B. ☐ Base Leg
C. ☐ Final Approach
D. ☐ Departure Leg or Exit
E. ☐ Not in Traffic Pattern
F. ☐ Unknown
G. ☐ Other, Specify _____

FAA Form 8020-17 (6-91)   Page 1

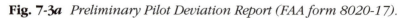

Fig. 7-3a *Preliminary Pilot Deviation Report (FAA form 8020-17).*

16. Aircraft Altitude When Deviation Detected:

A. └─┴─┘ , └─┴─┴─┘ Feet msl

B. ☐ Unknown

17. Transponder *(mark one)*:

A. ☐ Operating, With Altitude Reporting
B. ☐ Operating, Without Altitude Reporting
C. ☐ Not Functioning (broken or off)
D. ☐ No Transponder
E. ☐ Unknown

18. Was the Aircraft Equipped with TCAS?:

A. (1) ☐ Yes   (2) ☐ No   (3) ☐ Unknown
B. If Yes, Was TCAS Operating During Deviation?
   (1) ☐ Yes   (2) ☐ No   (3) ☐ Unknown
C. If Yes, Was TCAS Involved in Deviation?
   (1) ☐ Yes   (2) ☐ No   (3) ☐ Unknown
D. If Yes, Describe Involvement: _____
_____
_____

19. Fix or Facility Nearest Deviation *(complete one)*:

A. └─┴─┴─┘ VOR, TACAN or NDB ID
B. └─┴─┴─┘ Airport ID
C. └─┴─┴─┴─┘ Airway Intersection ID
D. ☐ Oceanic Airspace or Area Navigation (GPS, Loran, etc.)

20. Deviation Location in Respect to Item 19 *(complete A&B or C&D)*:

A. └─┴─┴─┘ Miles (nautical)
B. └─┴─┴─┘ Degrees (magnetic)

For Oceanic Airspace and Area Navigation Only:

C. └─┴─┘ ° └─┴─┘ '
   Latitude
D. └─┴─┘ ° └─┴─┘ '
   Longitude

21. Operational Control Area of Aircraft *(mark a maximum of three)*:

A. ☐ Class A Airspace
B. ☐ Class B Airspace
C. ☐ Class C Airspace
D. ☐ Class D Airspace
E. ☐ Class E Airspace
F. ☐ Class G Airspace
G. ☐ Special Use Airspace, Specify _____
_____
H. ☐ Within Terminal Radar Service Area
I. ☐ Towered Airport
J. ☐ Nontowered Airport
K. ☐ Unknown
L. ☐ Other, Specify_____

22. Location ID of Facility(ies) Providing Air Traffic Service During Deviation *(complete appropriate boxes)*:

A. └─┴─┴─┘ ARTCC
B. └─┴─┴─┘ TRACON
C. └─┴─┴─┘ RAPCON, RATCF, or ARAC
D. └─┴─┴─┘ ATCT

E. └─┴─┴─┘ AFSS or FSS
F. ☐ None
G. ☐ Unknown
H. ☐ Other, Specify _____

23. Preliminary Information Indicates the Air Deviation Type Was *(mark appropriate boxes)*:

A. ☐ ATC Altitude Clearance Deviation
B. ☐ ATC Course Clearance Deviation
C. ☐ Airspeed Clearance Violation
D. ☐ Airspace Clearance Violation
E. ☐ Flying VFR when IFR Required
F. ☐ Pilot Unqualified for Aircraft or Conditions

G. ☐ Required Aircraft Equipment Not Operating
H. ☐ Careless or Reckless Aircraft Operation
I. ☐ Unauthorized Low Level Flying
J. ☐ Missed Compulsory Reporting Point
K. ☐ Noncompliance with Other Regulations (specify FAR number[s]):
   (1) └─┴─┴─┘ . └─┴─┴─┘ (└─┘)   (2) └─┴─┴─┘ . └─┴─┴─┘ (└─┘)

24. Preliminary Information Indicates the Airspace Violation Was of *(mark one)*:

A. ☐ Class A Airspace
B. ☐ Class B Airspace
C. ☐ Class C Airspace
D. ☐ Class D Airspace
E. ☐ Class E Airspace

F. ☐ Special Use Airspace, Specify _____
G. ☐ None
H. ☐ Unknown
I. ☐ Other, Specify _____

25. If ATC Altitude or Course Clearance Deviation, Maximum Deviation Was:

☐ No Clearance Deviation

A. └─┴─┘ , └─┴─┴─┘ Feet, Vertical   or   ☐ Unknown
B. └─┴─┘ , └─┴─┴─┘ Feet, Horizontal
or └─┴─┘ . └─┴─┘ Miles (nautical), Horizontal   or   ☐ Unknown

26. If There Was Loss of Separation, Closest Proximity Was:

☐ No Loss of Separation

A. └─┴─┘ , └─┴─┴─┘ Feet, Vertical   or   ☐ Unknown
B. └─┴─┘ , └─┴─┴─┘ Feet, Horizontal
or └─┴─┘ . └─┴─┘ Miles (nautical), Horizontal   or   ☐ Unknown
C. └─┴─┘ Minutes, Longitudinal   or   ☐ Unknown

27. Other Reports Filed or To Be Filed *(mark appropriate boxes and complete)*:

A. ☐ Incident Report (FAA Form 8020-11), Specify No(s).
B. ☐ Preliminary Near Midair Collision Report (FAA Form 8020-21), Specify No(s). _____
C. ☐ Preliminary Operational Error/Deviation Report (FAA Form 7210-2.1), Specify No(s). _____
D. ☐ Other (including TCAS), Specify _____
E. ☐ None

28. Brief Description of Deviation and Comments *(comments optional)*:

_____
_____
_____
_____
_____

FAA Form 8020-17 (3-95) Supersedes Previous Edition   Page 2   NSB: 0052-00-899-0001

Fig. 7-3*b* *Preliminary Pilot Deviation Report (FAA form 8020-17). (Continued).*

		Incident Report Number							

**INVESTIGATION OF
PILOT DEVIATION REPORT**

P

Complete and distribute within 90 days of a reported pilot deviation according to instructions on page 3. Complete all items. Use the same incident report number as on the corresponding FAA Form 8020-17, "Preliminary Pilot Deviation Report." Any corrections to FAA Form 8020-17 should be reported in Item 17 of this form. Complete the form by hand or typewriter.

1. Date, Time, and Location of Deviation:

A. Date (Coordinated Universal Time-UTC)

M M D D Y Y

B. UTC Time

C. Local Time

D. Nearest City or Town and State

2. Pilot Information:

A. Name and Address

Name (first, middle, last)

Address

City   State or Country   ZIP

B. Home Base

C. Telephone Number

_____ - _____ - _____

D. Pilot Certificate No. (or enter "MILITARY")

E. Date of Birth

M M D D Y Y

3. Pilot Hours (if hours unavailable, estimate):

A. Total, All Aircraft _____ hours

B. Total, Make & Model in Deviation _____ hours

C. Last 90 Days, All Aircraft _____ hours

D. Last 90 Days, Make & Model in Deviation _____ hours

E. Duty Time, Last 24 Hours (includes Item 3F) _____ hours

F. Flight Time, Last 24 Hours _____ hours

G. Flight Time, Leg At Time of Deviation _____ . ____ hours

4. Pilot and Medical Certificate(s):

A. Pilot Certificate(s) (mark appropriate boxes):

(1) ☐ Student   (5) ☐ Airline Transport   (9) ☐ None
(2) ☐ Recreational   (6) ☐ Flight Instructor   (10) ☐ Unknown
(3) ☐ Private   (7) ☐ Military   (11) ☐ Other, Specify _____
(4) ☐ Commercial   (8) ☐ Foreign Pilot   _____

B. Medical Certificate(s) (mark appropriate boxes):

(1) ☐ First Class   (4) ☐ Special Issuance, Specify Type   (6) ☐ Out of Date
(2) ☐ Second Class   _____   (7) ☐ Unknown
(3) ☐ Third Class   (5) ☐ Self Certification   (8) ☐ None Required, Specify Reason _____

C. Date of Last Medical M M D D Y Y

5. Pilot Rating(s) (mark appropriate boxes):

A. ☐ Single Engine Land   F. ☐ Glider
B. ☐ Multiengine Land   G. ☐ Lighter-than-air
C. ☐ Single Engine Sea   H. ☐ None
D. ☐ Multiengine Sea   I. ☐ Unknown
E. ☐ Rotorcraft   J. ☐ Other, Specify _____

6. Pilot Instrument Rating (mark one):

A. ☐ Current
B. ☐ Not Current
C. ☐ None
D. ☐ Unknown

7. Prior Enforcement Actions Against Pilot (mark one):

A. ☐ One or More
B. ☐ None
C. ☐ Unknown

8. Date(s) of Pilot Checks and Tests (specify those within last two years, MM/DD/YY):

A. Flight Review

B. Proficiency

C. Competency Flight

D. Simulator

E. Route Check

F. Instrument Currency or Instrument Rating Flight Test

G. Airline Transport Pilot Flight Test

H. Flight Test (private, commercial, or flight instruction)

I. Other, Specify

FAA Form 8020-18 (3-95) Supersedes Previous Edition   Page 1   NSN: 0052-00-899-1001

Fig. 7-4a *Investigation of Pilot Devition Report (FAA form 8020-18).*

9. Aircraft Information:

A. Registration (N) No. [_ _ _ _ _ _]

B. Flight No. or Call Sign (if applicable)

C. Make   D. Model

E. Aircraft Type (mark one):
- (1) ☐ Single Engine Land   (5) ☐ Rotorcraft
- (2) ☐ Multiengine Land   (6) ☐ Other, Specify _____
- (3) ☐ Single Engine Sea
- (4) ☐ Multiengine Sea

10. Type of Operation at Time of Deviation (mark one):
- A. ☐ U.S. Air Carrier (14 CFR 121 or 125)
- B. ☐ Foreign Air Carrier (14 CFR 129)
- C. ☐ Commuter (14 CFR 135)
- D. ☐ Air Taxi (14 CFR 135)
- E. ☐ General Aviation (14 CFR 91)
- F. ☐ Public (governmental)
- G. ☐ U.S. Military, Specify Service _____
- H. ☐ Unknown
- I. ☐ Other, Specify _____

11. Aircraft Operator Information
(complete, or mark box if General Aviation):   ☐ General Aviation

A. Name and Address

Full Name _____

Address _____

City _____ State or Country _____ ZIP _____

B. Telephone Number [_ _] - [_ _] - [_ _ _]

C. Certificate Number [_ _ _ _]

12. Flight Information:

A. Departure Airport ID [_ _ _ _]

B. Destination Airport ID [_ _ _ _]

C. Local Flight:
(1) ☐ Yes   (2) ☐ No   (3) ☐ Unknown

D. First Flight of Day for Pilot:
(1) ☐ Yes   (2) ☐ No   (3) ☐ Unknown

13. Weather Contributed to Pilot Deviation (mark appropriate boxes):

- A. ☐ Pilot Received Inaccurate Weather Data
- B. ☐ Avoidance of Weather
- C. ☐ Flying Visual Flight Rules (VFR) in Instrument Conditions
- D. ☐ Unknown
- E. ☐ Other, Specify _____
- F. ☐ None of the Above, Weather Not a Factor

14. Aircraft Equipment Malfunction(s) Contributed to Pilot Deviation (mark appropriate boxes):

- A. ☐ Communication
- B. ☐ Transponder
- C. ☐ Navigation, Excluding Autopilot
- D. ☐ Autopilot
- E. ☐ Altimeter
- F. ☐ Unknown
- G. ☐ Other, Specify _____
- H. ☐ None of the Above, Equipment Malfunction Not a Factor

15. Investigation Indicates the Pilot Lacked or Had Inadequate Knowledge or Experience With (mark appropriate boxes):

- A. ☐ Aircraft
- B. ☐ Avionics
- C. ☐ ATC Procedures
- D. ☐ ATC Terminology or Phraseology
- E. ☐ English Language
- F. ☐ Preflight Planning
- G. ☐ Crew Coordination
- H. ☐ Weather
- I. ☐ Airport
- J. ☐ Current Charts and Approach Plates
- K. ☐ Unknown
- L. ☐ Other, Specify _____
- M. ☐ None of the Above

16. Investigation Indicates the Pilot Was (mark appropriate boxes):

- A. ☐ Overworked
- B. ☐ Distracted, Specify _____
- C. ☐ Fatigued
- D. ☐ Actively Scanning
- E. ☐ Not Actively Scanning
- F. ☐ Unable to Locate Traffic, Even With Traffic Advisory
- G. ☐ Disoriented or Lost
- H. ☐ Sick, Specify _____
- I. ☐ Not Following ATC Instructions, Specify _____
- J. ☐ Operating in Class A, B, C, or D Airspace Without Required Communication or Authorization
- K. ☐ Operating With Transponder Off
- L. ☐ Responding to TCAS Resolution Advisory
- M. ☐ Unknown
- N. ☐ Other, Specify _____
- O. ☐ None of the Above

17. Corrections and Additions to FAA Form 8020-17 (specify item number and new information or mark box): ☐ FAA Form 8020-17 is complete and accurate.

_____

Fig. 7-4b *Investigation of Pilot Deviation Report (FAA form 8020-18). (Continued).*

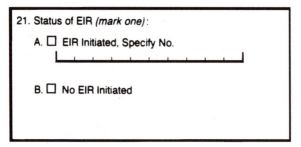

Fig. 7-5 *The box on the last page of form 8020-18 that records the investigator's decision.*

I work with believe that if they declare an emergency, they are going to be fined, lose their house, and be audited by the IRS. Those fears are truly unfounded, but the FAA does a poor job of public relations in this area.

Remember what the FAA inspector said: "Enforcement actions can only be taken against people, not against airplanes, so if the airplane breaks, no action (penalty) can be taken." That is good to know, but self-confidence also plays a role in the pilot's reluctance to declare an emergency. Looking back on emergency situations, some pilots were not sure about the regulations that might have pertained to their situation. Some did not know enough about the rules to know for sure if they violated one or not. Because they were not sure, they did not want to call attention to themselves by declaring an emergency for fear of exposing their lack of knowledge. Often it came down to a choice between the threat of danger and saving face. Saving face is the same as protecting that famous pilot ego. There is nothing wrong with the pilot's ego—I have a pretty big one myself. But you must "be, not just seem." So pilots have another choice. They can continually battle with this conflict between really knowing what's up or acting like they do when they don't, or they can hit the books, get refresher training, and start earning the right to walk around with the pilot's ego. Confident pilots get their confidence from their knowledge, not arrogance. So it is a cycle, greater knowledge is greater confidence. And the confident pilot knows that declaring an emergency is an option that can be freely used. Playing the emergency card, when it is needed, can make the flight safer. As pilot in command, your job is to maintain the highest level of safety, so keep your emergency card ready and do not be afraid to play it.

Pattern mismatches

The *U.S.S. Vincennes* was on routine patrol in the Persian Gulf when its radar detected an incoming aircraft. What happened next has become a classic example of a pattern mismatch that led to the shootdown of a civilian airliner. The ship's commander began to believe that the approaching airplane was hostile. This conception was built by confusing identification signals and misinformation about the aircraft's altitude. At one point the radar indicated that the aircraft was diving for the ship when in fact the airliner was climbing to cruising altitude. The commander, thinking that his ship would soon be under attack, took on a course of actions that would have properly matched a hostile threat. The commander followed the protocol perfectly: Warn off the aircraft and if the warning is not taken, shoot down the aircraft. That is exactly what the commander did. It was not until later that it was discovered that a terrible mistake had been made. The tragedy was not caused by the protocol, but by a pattern mismatch. The situation was tense and very time-sensitive. The commander was under pressure to act before the presumed attacking aircraft acted. The commander did everything by the book. But there was a problem. He had opened the wrong book.

Remember that in time-crunch situations there may be no time to deliberate the decision. You cannot call "time out" to weigh the pros and cons of a decision. Fast action is required while time still exists. The story of the civilian airliner shoot-down proves that these situations are not limited to airplane cockpits. Instant and critical decision making takes place in hospital emergency rooms every day. People face life and death decisions that must be made quickly if they work in law enforcement, fire fighting, air traffic control, nuclear powerplants, and countless military situations. I would imagine that the people in charge of launching nuclear ballistic missiles worry about this all the time. In these circumstances people tend to fall back on what they know, and this is why people with experience are the experts. People who have seen a similar situation before are more likely to remember what once worked to solve the problem and can draw on that memory to apply a fix to a current problem. This pattern-matching technique works, but you better make the correct match. If you choose a course of action that does not match the situation—disaster.

How can pilots ensure that they select the course of action that will match the situation and solve the problems at hand? Pilots must maintain an awareness of what is truly taking place in their environment. The ship commander would have never shot down an airliner had he known it was an airliner and of no threat. But the commander did not have a true understanding of the situation. The moment you lose touch (or never gain touch) with the circumstances that surround you, your choices are bound to be flawed. Examples of general aviation pilots in situations of pattern mismatches are detailed in the Appendix.

The proper pattern match chain appears to begin with the pilot's ability to control the airplane with a minimum of brain power. This leaves the pilot "space available" in the thought process to recognize and evaluate the problems that arise. The formula for a pattern mismatch chain appears to be the pilot's inability to think and fly simultaneously. A phrase that is common in flight training is "fly the airplane." This means that when faced with problems in flight, the first priority of the pilot is to maintain aircraft control, or fly the airplane. Some pilots can make aircraft control the priority but are unable to go much beyond that point. The first priority, aircraft control, completely saturates their thought process and robs them of the ability to use or understand other information that is presented. It all comes down to pilot knowledge and pilot proficiency. Knowledge allows the pilot to understand and match what goes with what, using associated groups. Proficiency allows the pilot time to see the big picture and apply the proper match.

8

Advanced Qualification Program

Special Federal Aviation Regulation (SFAR) 58 is a little-known part of the regulations. SFAR 58 allows air carriers to train pilots using proficiency and decision making rather than maneuvers for pilot evaluation. The regulation is called Advanced Qualification Program, or AQP. An AQP program shifts the pilot certification responsibilities away from the FAA and places the training responsibility on the airline that has an approved AQP program.

SFAR 58 1(a) reads, "This Special Federal Aviation Regulation provides for an alternate method for qualifying, training, certifying, and otherwise ensuring competency of crew members...." Today this "alternate method" can only be approved for pilots if they are already required to be trained under either Part 121 (the air carrier regulations) or Part 135 (the on-demand charter regulations).

An airline can have an approved AQP program as long as it meets the following requirements:

1 The training must include cockpit resource management (CRM).
2 The training must incorporate line-oriented evaluation (LOE), which is the logical way to test line-oriented flight training (LOFT).
3 The flight instructors and check airmen must undergo additional cockpit resource management training.
4 The airline must keep data for the FAA to use for performance assessment.

The SFAR specifically requires the pilots to be trained using real-world scenarios. SFAR 58. 7 (b) says "Approved training on and evaluation of skills and proficiency of each person being trained under AQP must use cockpit resource management skills and technical (piloting and other) skills in actual or simulated operations scenario." These scenarios must be played out in either approved flight training devices or flight simulators.

An approved program must have three curriculum sets: indoctrination, qualification, and continuing qualification. The indoctrination course is designed for newly hired employees of the company and covers company policies and general aeronautical knowledge. The qualification curriculum places a person into a specific duty position on a particular type of airplane. The continuing qualification curriculum establishes the cycle that ensures that those who have been trained remain proficient. Our general aviation equivalents are introductory flights, practical tests, and flight reviews.

Once an AQP program is approved, pilots are not held to any specific number of hours of experience but are instead tested on how they handle situations. The biggest difference between AQP and conventional training is that AQP is all about the mission not the maneuver. Pilots might be able to fly a great chandelle, but what will they do on a dark and stormy night? AQP does not completely eliminate the use of maneuvers, but it does make the assumption that maneuvers alone are inadequate.

A pilot might have 1000 flight hours, but that does not guarantee that the pilot can manage information, utilize all resources, and make decisions effectively. Flight hours are a gauge of experience, but no two pilots have had the same experience within their flight hours. Two pilots with 1000 hours have only the hours in common. The two pilots have faced completely different challenges during those 1000 hours. Therefore, flight hours are an uneven measure of experience. How then can flight hours be used as the sole assessment tool?

AQP eliminates flight hours as the assessment tool. Pilots qualify for a certificate or a type rating not when they accrue logbook time, but when they can deal with situations that would be encountered in the real world. Pilots in an AQP program are trained using LOFT and then tested using LOE. An LOE is a real-time flight from one airport to another that will involve several "event sets." An event set begins with an "event trigger." When the examiner introduces the trigger to the crew, the crew must react with a set of actions to meet the unusual occurrence.

Today, the airlines use the investigations from actual aircraft accidents and NASA forms to develop the event sets. When several event sets are used together, a full-blown setup scenario emerges.

Example 1: A fully loaded airliner departs from Denver en route to Phoenix. Coming out of Denver the crew is held at a lower than requested altitude of 10,000 feet. At the same time it is snowing and ATC gives the crew a complex change to their original route. The current heading is taking the airplane toward higher terrain. You can see that the crew is faced with several problems, but the low altitude is the trigger. Will the crew properly prioritize the combination of problems? The low altitude and the rising terrain must be the first problem. The snow and the route change should be second and third priorities. Will the crew deal with the problems in the safest order? The examiner observing the LOE will determine how effective the crew is and this will determine if it passes or fails.

Example 2: A crew is holding while awaiting an approach to Chicago. The weather is lowering to near Category III minimums. The airplane is approaching minimum fuel. Now what does the crew do? The trigger was the en route hold because that is what caused the fuel to be critical. The crew must now calculate how much fuel they can burn while holding before starting the approach. Then the crew should consider the possibility of a missed approach and the fuel requirement to make it to an alternate. The longer they must hold, the fewer alternate airports that are available. In this scenario the pilots must be able to (1) effectively fly a holding pattern entry and understand holding instructions, (2) make accurate fuel calculations, (3) determine the time it will take to attempt the instrument approach, and (4) select an alternate airport, taking into consideration weather and fuel range. Now it's one thing to do any one of these tasks, but it is quite another to handle all of these tasks at once when each action is interrelated to and will affect the other tasks. Putting it all together during a time crunch is AQP.

The examiner observing example 2 would be mildly interested in the hold entry "maneuver" but much more interested in the outcome of the entire scenario. Specifically, the examiner would watch for (1) the captain displaying authority, (2) the crew placing the fuel as the first priority, (3) the timing decision to hold, approach, or proceed to the alternate, and (4) the ability of each member of the crew to get involved in the decisions that were made. You can see that AQP is much more than testing a pilot on how he or she flies the instrument approach.

The AQP concept is also the real-world concept. But general aviation (GA) pilots have not been exposed to these AQP concepts on a large scale. When the general aviation pilots who were taught using flight-hour and maneuvers training came face to face with LOFT the results were shocking (see the appendix). They began crashing. When pilots who had been taught in an insulated training environment got thrown to the wolves in a real-world scenario, accidents happened. The GA pilots recognized the difference between how they were trained and this real-world approach. Some of their comments included:

- This taught me a different attitude.
- It was eye opening.
- This takes training in a different direction.
- I think it will change how I approach flying.
- This is a different world.

The GA pilots were aware that this was not something that they were used to, but nevertheless they seemed to understand the value of using real-world, or LOFT-type, training:

- Overall, this project will help aviation safety. Instructors should use real-life situations in training and recurrent training.
- It was *extremely* beneficial for me. I'd like more real-life situations in my training.
- I wish something of this nature had been a part of earlier training.
- I wish I had a regular session with similar real-world scenarios at least every 6 months. It should be required for recurrency.
- Everyone should have this opportunity.

When the GA pilots were asked if they thought the real-world method should be used in initial pilot training, 98.2 percent said Yes. When asked if LOFT-type training should be a part of recurrency training, the GA pilots again spoke as one when 94.7 percent said Yes. These pilots were not afraid to try something new if they thought it would make them safer pilots.

So here are the facts that we know of: The FAA now approves mission as opposed to maneuvers training with the airlines. The GA pilots who were exposed to mission-type training made big improvements (see the Appendix) in decision making. The GA pilots were greatly challenged by the real-world method, which was new to them, but nevertheless resoundingly asked for more and recommended it for other GA pilots. So I think it is time for a GA advanced qualifications program.

I am not advocating the complete abandonment of maneuvers training. Of course pilots must learn to land in a crosswind, recover from stalls, and fly instrument approaches, but we should do more. We should do better. We should incorporate decision training and LOFT into our GA flight training and make it as normal as learning to taxi.

The current FAA regulations say that a person is qualified to have a private pilot certificate after 40 hours of flight experience. Will any two pilots be exposed to exactly the same experience during those 40 hours? No, so why is there a one-size-fits-all regulation? Do these 40 hours guarantee that the pilot will be an effective decision maker in addition to a proficient maneuver maker? No single number should be considered the magic number. A person should become a pilot on the day that he or she can display safe aircraft manipulation and safe decision-making abilities together. For some, this will take more than 40 hours; for others, it might take less time.

Of course, the hitch in this plan is in the pilot testing. The FAA will never use a more practical approach unless it can be assured that the process is just as effective as the one they have now. The FAA already charges its inspectors and designated examiners to include CRM into every checkride. The private pilot practical test standard (PTS) says "CRM is not a single task, it is a set of skill competencies that must be evident in all tasks in the PTS as applied to either single pilot or a crew operation." But CRM is only one part of the real-world method of flight training. The FAA should do more. The FAA should do better.

Conducting a standardized test is a challenge for any organization. How can anyone be sure that a person given a private pilot checkride in Modesto is meeting the same standard as a person taking the same checkride in Valdosta? Many years ago the FAA replaced the old flight test guides with the PTS. The PTS has removed some of the variability and flexibility from the tests, and it has become more standard, but any instructor will tell you that no two examiners test alike and no two tests consist of exactly the same items. Incorporating more decision-based tests and training examiners to administer such tests would require some effort, and it would also require a partnership. The current airline AQP programs rely heavily on the airlines themselves to train and evaluate. Likewise, the FAA would have to rely heavily on flight instructors to do the preliminary decision training and evaluation. This means that CFIs must get on board, and the way I see it, they must get on board first.

Excellent flight instruction depends on many factors. But it is the flight instructor who will make the biggest impact on quality. It really does not matter whether the school is part 141 FAA approved, or part 61. It really does not matter whether the airplanes are high wing or low wing. It does not matter whether the lessons are conducted at a busy Class B airport or a grass strip. What matters is the flight instructor and how he or she gets the message across and how the student works to get the message. The FAA might never accept a general aviation AQP program, but in the meantime flight instructors should do a better job of training with the use of naturalistic (real-world) decision-making techniques.

The SFAR 58 now allows those enrolled in an approved AQP program to receive their commercial or airline transport pilot certificates without the conventional checkride. The certificate is awarded after the completion of the AQP curriculum. Once again the final evaluation for completion includes real-world scenarios. SFAR 58, 8(c) states "[to pass the course] an applicant must show competence in required technical knowledge and skills (e.g., piloting) and cockpit resource management knowledge and skills in scenarios that test both types of knowledge and skills together." That last part "test both types of knowledge and skills together" is the essence of the whole idea. Can a person operate the machine (airplane) and at the same time use resources and make decisions to ensure a safe flight and do this all at the same time? Many general aviation pilots cannot, but don't blame them completely because they were never actually trained to do it in the first place.

AQP itself is a continuum. For years airlines had no accidents because of errors in pilot technique. The accidents that did take place were due to errors in judgment and decision making. So AQP was designed to fight that problem. But has AQP swung too far? Are pilots now less skilled in the actual manipulation of the machine because AQP placed emphasis on crew coordination and decision making? AQP within the airlines must itself strike a balance between maneuvers and mission. This will mean even more training: part on repetitive drill and practice of maneuvers and emergency procedures and part on CRM and decision making. The cost of safety will go up.

As for general aviation, AQP may never become a reality. In the meantime pilots and instructors are not prevented from using AQP techniques inside their normal training and recurrency. The next chapters will be a guide for pilots and CFIs to incorporate the AQP concept into everyday flight training.

9

Decision Training
for Student Pilots

This chapter is filled with examples on how to use real-world decisions as a part of everyday flight training. It is for both the student and the instructor. Instructors must be storytellers and students must be riddle solvers. It will take creativity on both sides, but the result will be pilots who are better prepared to deal with situations in the flight environment.

In this chapter we look at the logical divisions of the training that lead to the private pilot certificate. Also, we look at the FAA PTS and how real-world, not just how-to-pass-the-test, instruction can aid in preparing for the checkride and beyond. The references to the PTS that are cited are not included to be critical of the standard but to point out that the PTS is very maneuver-based and to offer incentives for pilots, instructors, and examiners to expand the PTS to include scenario-based training.

In flight training the building-block method works well. This is also known as the you-must-walk-before-you-can-run method. A simpler skill is taught so that it can be used to facilitate the learning of a more complex skill. The skills build one on the other with mastery being the goal. I think of a private pilot course as shown in Fig. 9-1. This picture has all the major building blocks that are required for a person to become a private pilot, and they are prioritized from the ground up. Pilots cannot be expected to fly the airplane solo until they have been taught to fly the traffic pattern, and they cannot be expected to fly a solo cross-country unless first they had dual cross-country instruction. So the blocks are prioritized and stacked so that a firm foundation is built. Each block has an entrance and an exit. The entrance is the lesson objective, or a statement of what it is you want to get accomplished in this block. The exit is the completion standard, or what the person must be able to do to move on to the next block. Meeting the standard of one block means it is time to move up to the next block. When someone arrives in the next block, he or she will be using the knowledge and skills learned in the previous block to work toward the completion standard of the new block. Notice that the number of flight hours is never shown here. One student might need five 1-hour lessons to meet the standard of a particular block, while another student gets it done in 3. With this method, the use of minimum flight hours is out and standards are in.

Today the completion standards are mostly maneuvers that a person can perform. Once someone can manipulate the airplane to a particular skill level, he or she moves on. Learning to be the pilot in command takes more than machine manipulation, however. This whole book centers around this fact: We have not done the best job possible in teaching people to be pilot in command. If we want to continue with the past results, we should continue with the past methods. But that is no longer good enough. We need new methods that will keep the best from the past but incorporate what teaches people to become decision makers, not just manipulators. So the new completion standards must include decisions. The real-world offers the pilot an unlimited number of decision circumstances, so we should be placing would-be pilots into circumstances where decisions are required and let them practice making them. This real-world element can be injected into any flight training situation if the student and instructor use a little creativity.

The following are some examples of how the real-world element can be a part of everyday flight training, with the goal being to not only train to be a pilot but to teach a person how to become a pilot in command. These examples are certainly not the last word or the only way to do it. Pilot students and instructors are every bit as creative as I and then some. But let these examples get you thinking.

The four fundamentals

Climbs, descents, turns, and straight-and-level flight are the basic fundamental maneuvers that an airplane can do. No matter how many airplanes you fly, no matter how many flight hours you acquire, even if you fly the space shuttle, there are only four maneuvers that airplanes can do. We sometimes will combine these maneuvers, but even the most complicated aerobatic maneuver is still a combination of the four fundamentals.

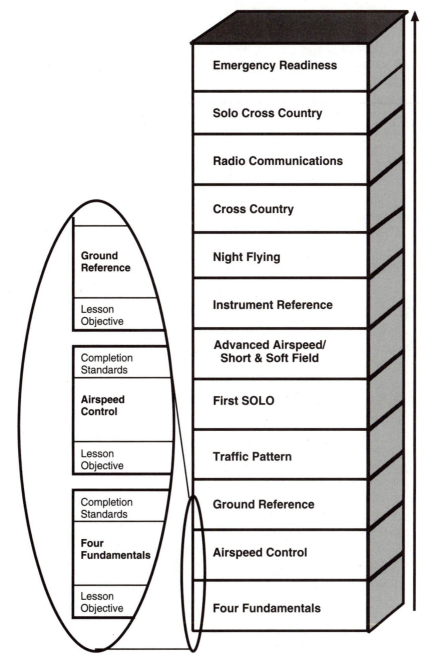

Fig. 9-1 *Private pilot building blocks.*

Scenario: You are a student pilot flying back from the practice area to the traffic pattern. You are level at 2000 feet. Without warning you see a flock of large birds just in front of you. It appears that you will hit at least one of these birds if you don't do something. What should you do?

Well, you probably will have to take some evasive action to avoid a midair collision with the flock. Which of the four fundamentals might you need? Probably all of them. You might even need to combine a few of the four just to get through. A quick climbing right turn or a left descending turn might do it. The actual decision of what combination to use would be decided by the actual circumstance, but if the use of the four fundamentals is ever needed, as they would be in this scenario, we better practice the combinations. The four fundamentals are simple maneuvers, and this scenario is a simple situation, but maneuvers and decisions were combined.

In the past the instructor would say, "Turn right to heading 180. Now climb to 2500 feet. Now turn to 360 and climb to 3000 feet at the same time." The fundamentals are being taught, but no decisions are being made.

Airspeed control

When flying airplanes, we must maintain a forward speed so that we always have airflow over the wings. After all, airplanes are not hot air balloons. They cannot just hang in the air. But even in light airplanes changing speed is often necessary.

Scenario: A pilot is approaching an airport. The airplane is 1000 feet above the ground and in the airport's traffic pattern. The airplane must do two things to land. It must descend to the ground elevation and also reduce its speed so that the wings will stop flying. How will the pilot get the airplane to both fly "downhill" and, at the same time, slow down?

Usually when something moves downhill, it accelerates and the speed increases, but if that happens in this case, the airplane will never land because it would arrive at the runway going too fast. What do you do?

This question incorporates many maneuvers and decisions. How do the wings create lift? How will lowering the flaps affect the airplane?

What controls the airplane's speed anyway: power or pitch? When should power be reduced and/or pitch increased?

In the past, the instructor would say, "Give me 60 knots on this heading. Now keep 60 knots, but turn right to 090. OK, now hold 090, but descend to 3000 feet." The airplane is changing speeds and changing altitudes, but there is no apparent reason for doing this. I thought airplanes were supposed to go fast, so why are we deliberately flying them slowly? A maneuver is being taught, but there is no reason to do it without including a situation where this maneuver might be necessary.

The private pilot PTS requires maneuvering during slow flight. The objective is to determine that the applicant

1 Exhibits knowledge of the elements related to maneuvering during slow flight
2 Selects an entry altitude that will allow the task to be completed no lower than 1500 feet above ground level (AGL) or the recommended altitude, whichever is higher
3 Stabilizes the airspeed at 1.2 Vs1, with a +10- and –5-knot tolerance
4 Accomplishes coordinated straight-and-level flight and level turns, at bank angles and in configurations, as specified by the examiner
5 Accomplishes coordinated climbs and descents, straight and turning at bank angles and in configurations, as specified by the examiner
6 Divides attention between airplane control and orientation
7 Maintains the specified altitude, with a ±100-foot tolerance. A ±10° heading tolerance and a +10- and −5-knot airspeed tolerance
8 Maintains the specified angle of bank, not to exceed 30° in level flight +0/−10°; maintains the specific angle of bank, not to exceed 20° in climbing or descending flight, +0/−10°; rolls out on the specified heading within ±10°; and levels off from climbs and descents within ±100 feet

You can see that the PTS is speaking only to the performance of a maneuver, and no effort is given to understanding when this situation might arise and what decisions might surround it. There are opportunities for a creative pilot examiner to include scenarios and decisions, but they are not required. This maneuver is very important for piloting skills and every pilot should be able to meet these standards, but the current requirements do not require the inclusion of decision skills.

Ground reference maneuvers

I was giving a stage check to a pilot who had just soloed the day before and I asked him to do the "turn around a point" maneuver. There was at least a 12-knot wind blowing, which offered a challenge for that maneuver, but nevertheless this student pilot flew the maneuver perfectly. I was impressed. I told the student that he had done a really good job outsmarting the wind and then asked, "Why do we do those maneuvers anyway?" The student said proudly, "Well it's a practice for later when I'm working on my instrument rating, in case I have to do a holding pattern." That really was not the answer I expected. I had a talk with his flight instructor later. It does no good to teach something if the learner sees no use in the lesson.

Scenario: Two airplanes are both flying downwind in a traffic pattern (Fig. 9-2*a*). There is a crosswind present. Airplane 1 continues the downwind leg without adjusting for the crosswind. Airplane 2 continues downwind but crabs into the wind slightly (Fig. 9-2*b*). The wind has drifted airplane 1 out away from the airport so that when airplane 1 turns on the base leg, a conflict arises with airplane 2 (Fig. 9-2*c*). What should happen next time to avoid this problem?

Pilots cannot actually see the wind, but they must fly their airplanes as if they do. Pilots must manipulate the airplane while flying straight and while in turns so that their track across the ground seems unaffected by wind drift. If pilots do not do this, their ground tracks will be somewhat out of control, and in the traffic pattern that can be dangerous. So pilots learn to fly parallel to a course even though there is a crosswind by using a "crab," or "wind correction angle." That maneuver is called a parallel course. Pilots also learn to vary the bank angle to avoid wind drift while in turns. That maneuver is called the turn around a point. Both these maneuvers are classified as ground reference maneuvers because the pilot must fly the airplane through a wind by reference to some object on the ground.

Placing the ground reference maneuvers in a scenario, especially a scenario that could involve a midair collision, forces the pilot into a decision to use the maneuver to solve the problem.

In the past, the instructor would say, "Fly a circle around that silo." The pilot then flies a circle around the silo but really does not know how this might ever be applied, and no decisions were required.

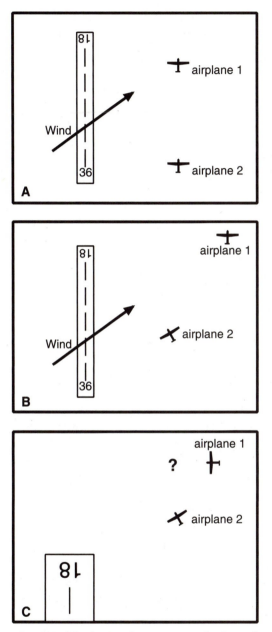

Fig. 9-2 *The hazardous potential of wind drift in the traffic pattern.*

The private pilot PTS includes the turn around a point as a task for use on the checkride. The objective is to determine that the applicant

1 Exhibits knowledge of the elements related to turns around a point
2 Determines the wind direction and speed
3 Selects the reference point with an emergency landing area within gliding distance
4 Plans the maneuver so as to enter at 600 to 1000 feet AGL, at an appropriate distance from the reference point, with the airplane headed downwind and the first turn to the left
5 Applied adequate wind-drift correction to track a constant radius circle around the selected reference point with a bank of approximately 45° at the steepest point in the turn
6 Divides attention between airplane control and the ground track and maintains coordinated flight
7 Completes two turns, exits at the point of entry at the same altitude and airspeed at which the maneuver was started, and reverses course as directed by the examiner
8 Maintains altitude ±100 feet and maintains airspeed ±10 knots

Why is the recommended entry altitude between 600 and 1000 feet? The FAA *Aeronautical Information Manual* (1999) says that "at most airports and military bases, traffic pattern altitudes for propeller driven aircraft generally extend from 600 feet to as high as 1500 feet above the ground." Pilots should practice their ground reference maneuvers at the same AGL altitude as their home airport's traffic pattern because ground reference maneuvers are just a traffic pattern warm-up. This fact is not stated in the PTS. According to the PTS you just do the maneuver.

Traffic pattern

To any student pilot the practice area is like the minor leagues is to a baseball player. It is the minor leagues when you learn your craft, yet the stakes are not all that high. To the student pilot the traffic pattern is the major leagues. The pattern is the show. In the traffic pattern you have a much bigger audience of people on the ground, other pilots in the air; even air traffic controllers can be among the spectators. And in the traffic pattern the stakes are much higher. In the practice area, it really did not matter if you were a little off altitude or a little blown off by the wind. But in the pattern with all the other airplanes, everything does matter because everything you do affects everyone else.

Scenario: You are on the downwind leg of the traffic pattern and are setting up for a normal landing. As you are just about to turn base, you see a large airplane making a straight-in approach to *your* runway (Fig. 9-3). What do you do now?

Should you extend the downwind leg or turn base? How will this approach need to be adjusted to compensate for a longer downwind? Will power, airspeed, and flaps settings be different this time? Is now a good time to introduce wake turbulence? Is the big airplane on a straight-in instrument approach, the location of which we should talk about? There are plenty of decisions to be made in this "event set."

In the past, the instructor would say, "This time let's fly out farther on downwind to see if you can do a longer final approach." The longer final maneuver was taught, and the student performed the maneuver well, but the entire time the student was thinking about the maneuver, not making decisions that would have led up to that maneuver.

The private pilot PTS requires the applicant to

1 Exhibit knowledge of the elements related to the traffic pattern. This shall include procedures at controlled and uncontrolled airports, runway incursion and collision avoidance, wake turbulence avoidance, and wind shear.
2 Comply with traffic pattern procedures.
3 Maintain proper spacing from other traffic.
4 Establish an appropriate distance from the runway, considering the possibility of an engine failure.

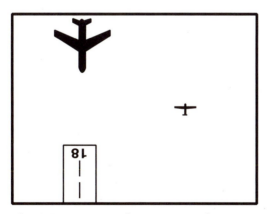

Fig. 9-3 *Large airplane on straight-in approach creates an event set for the smaller airplane.*

5 Correct for wind drift at maintain proper ground track.
6 Maintain orientation to the runway.
7 Maintain traffic pattern altitude within ±100 feet and maintain an appropriate airspeed within ±10 knots.
8 Complete the appropriate checklist.

Most midair collisions take place on VFR days in uncontrolled airport traffic patterns. This should provide all the incentive necessary to do a good job with traffic pattern scenarios.

First solo

One of the great adventures for every pilot is that first solo flight. As an instructor I will sit beside the student during those last dual landings before solo, and I will try to remain quiet. I will not offer suggestions. It is as if I am not there. Then I can say to the student, "I don't know if you noticed, but I didn't help you at all on those last landings. You have been doing it all yourself, so it really won't matter if I'm in here with you or not anymore." Then I let him or her solo thinking that in reality he or she had been soloing all along. The fact that I am not physically present should not change a thing (except the airplane's weight).

The first solo is not actually a new maneuver, but to the student pilot it is a mission. Decisions should be included with the student's first solo as well, always remembering that the instructor has the final call. An instructor might ask the student, "What do you think about this wind?" The student looks at the midfield wind sock and says, "Well it's kind of breezy, but it is right down the runway." The student recognizes that the wind is within his or her own tolerance. Later, after several dual takeoffs and landings to ensure consistency, the final decision is made to solo. The instructor makes this final decision, but the student is included in the process.

Advanced airspeed maneuvers

The short- and soft-field takeoffs and landings are actually more than they seem. The short- and soft-field takeoff and landing techniques, as the name implies, prepare the pilot to operate out of an airport that has a shorter than normal and/or nonhard surface runway, but it is more than that. To the student pilot these maneuvers are a demonstration of advanced airspeed control in a critical situation. Each of these maneuvers requires a very close tolerance of airspeed control, and the airspeed that is used is slower than normal. Combine a slow airspeed with a low proximity to the ground/runway and you have a critical situation. A false move here at 50 feet AGL will cause more of a hazard than a false move while in the practice area at 3000 feet AGL. This is really the first time that the student pilot learns precision flying.

Scenario: A pilot in a light airplane is flying to a busy controlled airport. There are two runways at this airport. One is long and is primarily used for air carrier traffic; the other is short and is used for general aviation traffic (Fig. 9-4). The controller tells the pilot, "You are cleared to land on runway 27, hold short of runway 36. Boeing 737 traffic on 5-mile final for runway 36." Is it possible to land on runway 27 and come to a complete stop before crossing runway 36? Must the light airplane pilot accept these instructions? What should the pilot do?

The pilot could ask the controller about the distance from the runway 27 threshold to the runway 36 intersection. The pilot could then combine the available landing distance information with the airplane's

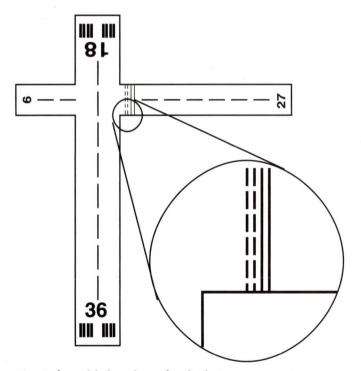

Fig. 9-4 *Hold short lines for the longer air carrier runway.*

short-field landing capability and make a decision to accept or reject the instructions.

Note: The Airline Pilot's Association recently rejected all land and hold short operations (LASHOs) when a general aviation airplane was approaching the crossing runway. They did this because they do not trust general aviation pilots. They clearly feel that our initial and recurrent training is inadequate to ensure that we will do the right thing each time. They have a low opinion of general aviation pilots and our decision-making skills. Maybe their opinion of us would improve if we saw the short-field landing not just as a maneuver but as a potentially dangerous scenario.

In the past, the instructor would say, "Let's see if you can land and make that first taxiway turnoff."

The private pilot PTS includes the short-field approach and landing. To meet the standards of the test the applicant must

1 Exhibit knowledge of the elements related to a short-field approach and landing
2 Consider the wind conditions, landing surface, and obstructions and select the most suitable touchdown point
3 Establish the recommended approach and landing configuration and airspeed and adjust pitch attitude and power as required
4 Maintain a stabilized approach and the recommended approach speed, or in its absence not more than 1.3 Vso, with a +10- and −5-knot tolerance with gust factor applied
5 Make smooth, timely, and correct control applications during the roundout and touchdown
6 Touch down smoothly at the approximate stalling speed, or within 200 feet beyond a specified point, with no wind drift and with the airplane's longitudinal axis aligned with and over the runway centerline
7 Apply brakes, as necessary, to stop in the shortest distance consistent with safety
8 Maintain crosswind correction and directional control throughout the approach and landing
9 Complete the appropriate checklist

All these PTS requirements never mention the possibility of conducting this maneuver while another airplane is bearing down on a collision course, but the scenario-based training does.

Instrument reference

You do not have to be an instrument flight instructor (CFII) to give instrument flight instruction. The only time a CFII certificate is required is when a person is training for the instrument rating. Initial instructors give instrument training all the time. In fact, it is a requirement that all student pilots receive instrument reference instruction.

Scenario: A VFR pilot is returning from the practice area to the home airport. The pilot had been practicing stalls and therefore was up at 4000 feet AGL. The sun was setting and the pilot, while enjoying the sunset, lost track of time. Without noticing it a thin and then a thick layer of clouds formed at a lower altitude, which is between the pilot and the airport. The pilot flies toward the airport only to discover that the clouds are covering the airport and now have covered the area behind as well. Now what?

What are the options for this pilot? Fly to another airport? Circle and hope the clouds go away? Fly through the clouds to the airport? Should the pilot have kept a better eye out for cloud cover in the first place? Decisions, decisions. I would hope that the pilot would have been sharp enough not to have let this happen in the first place. But if, despite all efforts, it happens anyway, I would hope the pilot could fly to another airport in VFR conditions. But in a worse-case situation, if the clouds layer was everywhere and it was getting dark, I guess an emergency descent through the clouds would be the last choice. I would only hope that the pilot had done some hood work before this flight and had practiced a slow rate-of-descent, no turns, let down. Just talking about this dangerous situation will scare a wise pilot into keeping one eye on the clouds while enjoying that sunset.

Confronting pilots with this story will show them a greater need for instrument reference maneuvers. I also hope they understand that instrument reference maneuvers for the VFR pilot are the last resort and that they should be vigilant to changing weather conditions.

In the past, the instructor said, "Here, put this hood on."

The student replied, "Why do we have to do this—that hood hurts my head and it's too hot."

"Because we have to get 3 hours under the hood for the checkride."

With regard to constant airspeed descents the private pilot PTS requires the applicant to

1 Exhibit knowledge of the elements related to attitude instrument flying during straight, constant airspeed descents
2 Establish the descent configuration specified by the examiner
3 Transition to the descent pitch attitude and power setting on an assigned heading using proper instrument cross-check and interpretation, and coordinated control application
4 Demonstrate descents solely by reference to instruments at a constant airspeed to specific altitudes in straight flight
5 Level off at the assigned altitude and maintain that altitude within ±200 feet; maintain the heading within ±20°; and maintain airspeed within ±10 knots

Again, all pilots should be able to do this, but the PTS does not present the life-and-death reason for doing it, which scenario-based training does.

Night instruction

My first instructor told me prior to my first night lesson that the airplane does not know that it is dark outside. The airplane flies exactly the same; it's the human body that acts differently at night. Our eyes, our oxygen intake, and our fatigue affect our bodies differently at night. Night flying can be wonderful. But it has dangers that the pilot must respect.

Scenario: A student pilot flies on a solo cross-country flight 100 miles from the home airport. When the pilot lands and shuts down the engine, he or she inadvertently leaves the master switch on. One hour later the pilot returns to the airplane to discover that the battery is dead. The pilot spends another half hour arranging for a power cart to be taken to the airplane for a jump start. Finally, the pilot gets back in the air and on the way home but now is at least 1 hour behind schedule and the sun is going down. The pilot has never made a solo night landing but faces one now. What should the pilot do?

Land somewhere else before it gets dark? Continue on because you have to be at work first thing in the morning and night landings are no big deal anyway? If a night landing is attempted, and while on final approach you see that the runway lights appear to "twinkle," what is happening and what should you do? What if you turn on the landing light and nothing happens? Was it burned out all along, but you never checked it because you never thought you would be flying at night?

The pilot in this story has many concerns to weigh. Landing somewhere else before dark is probably best, but what if there is no place else you can get to before dark? At night "twinkling" lights usually mean that tree branches or power lines are passing between your eyes and the lights. Are you too low? Should you go around? You should check the landing light during preflight, but can you land an airplane at night without a landing light? Ask all these questions and understand the reason for procedures at night and gain a respect for night flight.

In the past, the instructor would say, "C'mon, it's dark enough. I have been here all day, you know. The book says you need 10 takeoffs and landings at night to get your ticket. Did you bring a flashlight?"

The private pilot PTS does not require that any part of the checkride be conducted at night, but the examiner is to orally evaluate the applicant to determine that they can explain

1 Physiological aspects of night flying, including the effects of changing light conditions, coping with illusions, and how the pilot's physical condition affects visual acuity
2 Lighting systems identifying airports, runways, taxiways, obstructions, and pilot controlled lighting
3 Airplane lighting systems
4 Personal equipment essential for night flight
5 Night orientation, navigation, and chart-reading techniques
6 Safety precautions and emergencies peculiar to night flying

And during the flight portion of the practical test the examiner will evaluate the applicant's ability to:

1 Exhibit knowledge of the elements related to night flight
2 Inspect the interior and exterior of the airplane with emphasis on those items essential for night flight
3 Taxi and accomplish the before-takeoff check, adhering to good operating practices for night conditions
4 Take off and climb with emphasis on visual references
5 Navigate and maintain orientation under VFR conditions
6 Approach, land, and taxi, adhering to good operating practices for night conditions
7 Complete all appropriate checklists

Night flying and night flight instruction can be fun and rewarding. Do not forget that it is a great learning experience that can be filled with decision situations.

Cross-country

Flying away from the friendly confines of your home airport offers another great flying challenge. It also offers an unlimited number of what-if scenarios. When the airlines use LOFT scenarios, they are always going somewhere to begin with. Creative instructors and inquisitive students can "war game" cross-countries forever. Here is just one: a true story.

Scenario: A pilot and three friends piled into a light airplane for a flight to a small airport to attend an auto race. This was the third consecutive year these friends had made this pilgrimage to this particular race. Almost, but not quite to the destination airport, the engine of the airplane quit. The fuel tanks were dry. The pilot made an emergency landing in a field just at dusk and except for a nasty black eye the pilot suffered, all were unhurt. The local newspaper was on the scene within minutes and interviewed the pilot. "What happened?" the reporter asked. "I just don't understand it," the pilot said, "we made this trip down here on a full tank of gas the last 2 years without any problem!"

What did the pilot forget? Could the wind have been different for this year's trip? Was there any way the pilot could have known whether or not he had a headwind to fly against? How could this have been handled differently?

This story has helped many of my students *decide* to plan checkpoints, keep track of their time between checkpoints, and calculate groundspeed while en route. If they discover that they are moving across the ground more slowly than they had planned, they will have time to decide whether or not an unplanned fuel stop will be necessary.

In the past, the instructor would say, "Before we go cross-country next week I want you to complete this worksheet of flight computer problems." Or they would say, "we always make this trip down there on a full tank of gas without any problem."

To become a private pilot by the practical test standards the examiner must determine that the applicant

1 Exhibits knowledge of the elements related to cross-country flight planning by presenting and explaining a preplanned VFR cross-country flight near the maximum range of the airplane, as previously assigned by the examiner. The final flight plan shall include real-time weather to the first fuel stop, with maximum allowable passenger and baggage loads.
2 Uses appropriate, current aeronautical charts.
3 Plots a course for the intended route of flight.
4 Identifies airspace, obstructions, and terrain features.
5 Selects easily identifiable en route checkpoints.
6 Selects the most favorable altitudes, considering weather conditions and equipment capabilities.
7 Computes headings, flight time, and fuel requirements.
8 Selects appropriate navigation systems/facilities and communication frequencies.
9 Confirms the availability of alternate airports.
10 Extracts and records pertinent information from NOTAMs, the *airport/facility directory,* and other flight publications.
11 Completes a navigation log and simulates filing a VFR flight plan.

On the flight portion of the practical test the examiner will test the applicant for pilotage and dead reckoning skills, the use of radar and radio navigation systems, diverting to an unplanned alternate, and lost procedures. Of all the parts of the practical test, this one is the most mission-based and least maneuver-based. Most examiners do a good job of creating a scenario around these tasks. This proves that testing can combine decision making inside scenarios that are used for general aviation pilot evaluation.

Radio communications

Students who learn to fly at controlled airports never seem to have this problem, but students who learn at uncontrolled airports seem to develop a fear of the radio. All pilots must eventually get past the stage fright that is associated with the radio and get to the point where communications become conversational. I don't mean that you should chat with the controller during heavy workload times. But at the same time you should not feel inhibited or afraid of saying something wrong.

Scenario: A student pilot is flying into a busy Class C airport. The pilot makes the initial call-up, is assigned a transponder code, and is radar identified. The student is told to fly a heading for "sequencing." The pilot flies the heading and listens to abundant rapid fire radio communications. The pilot looks ahead and sees two very tall antennae. The antennae look as if they are on the assigned heading and at about the same altitude as the airplane is flying. The radio communication taking place on the frequency is without break. It would seem very hard to get a word in, but those towers are getting closer, and the controller has not called the airplane's number in a long time. Has the controller forgotten about the pilot? What should the pilot do?

Should the pilot wait it out and hope the controller calls with a turn? Maybe the towers are not as tall as they look. What if the pilot was given a heading that took the plane near the towers, but instead of rapid radio conversation, the pilot heard nothing at all on the approach frequency? Has the correct frequency been selected? Are the switches on the audio panel in the correct position? Do you remember the light gun signals? These events would certainly put the pilot into a decision situation. What should the pilot do? What would you do?

In the past, the instructor would say, "I made you a copy of some things to say to the controller. Just read from this script and you will do fine." Of course, the first thing the controller says is not on the script and the student says, "Ah...."

The private pilot PTS includes the use of light gun signals with radio communications evaluation. On the test the examiner's objective is to determine that the applicant

1 Exhibits knowledge of the elements related to radio communications and ATC light signals. This shall include radio failure procedures.
2 Selects appropriate frequencies.
3 Transmits using recommended phraseology.
4 Acknowledges radio communications and complies with instructions.
5 Uses prescribed procedures following radio communications failure.
6 Interprets and complies with ATC light signals.

Becoming familiar with the radio is one of piloting's greatest hurdles. The best way to clear this hurdle is to listen and then attempt to make it conversational not scripted.

Solo cross-country

Pilots should practice cross-country planning, even on days when it does not look like a flight is possible due to weather. Call and get a weather briefing anyway. Get the wind and temperatures aloft so that you can calculate the groundspeed and fuel requirements. Instructors should have the students make the go/no go decision. As an instructor I always go behind the student and get my own weather briefing. I secretly make the decision as to whether or not I want the student pilot going solo cross-country, but I keep my decision to myself. I wait until the student reports back to me about the weather, and I hear the decision that the student has arrived at. Most of the time, our decisions will match. If they do not match, this will start a discussion about decision making.

Scenario: A student pilot is returning home on the last leg of a solo cross-country. The weather has been perfect. The student's first checkpoint is a power line and a road, but when the time to be over the checkpoint runs out, the checkpoint is nowhere in sight. The student descends a little lower to get a better look. The student had been using a VOR station as a cross-check, but now it seems to have gone off the air. The student continues, but now nothing looks familiar, and soon the student is completely turned around. What should the student do?

The student first could have selected a better checkpoint than a power line and a road. There are so many power lines and roads that are not on the chart, singling out a specific pair would be tough. Use an interstate highway, river, town, or anything that is easier to see. Should the pilot have descended? Probably not. The VOR going off the air could have been nothing more than terrain blocking the signal. A climb will turn it on again. Should the pilot complete a VOR cross-check to find position? Should the pilot call ATC or FSS? Should the pilot circle in that area and hope to recognize something? Placing these decisions before the pilot before they take place makes the act of making these decisions commonplace.

In the past, the instructor would say, "Do you have your chart? Do you have your weather? Let me sign off your logbook, and here is the number I'll be at if you need me."

Emergency readiness

The research shows that pilots who have seen a problem before and been able to deal with it will be more prepared to handle a similar problem the next time. Some emergencies can effectively and safely be simulated in the airplane. We have all had an instructor pull an engine on us when we least expected it. These are "get down" emergencies where most decisions are made for the pilot. If an engine just quits, the airplane will start down, and there is no decision the pilot can make that will stop that from happening. Worse yet, if a fire erupts in flight, there will be no choice other than to go down fast.

Before a pilot can solo for the first time he or she must practice the scenario of engine failure. FAR 61.87(d)(11) and (13) require student pilots to undergo emergency training before they fly solo. Part 11 refers to emergency procedures and equipment malfunctions and part 13 requires the student to do approaches to a landing area with simulated engine malfunctions. Likewise private pilot applicants must show evidence that they are proficient in emergency procedures to be eligible to take the checkride [FAR 61.107(b)(x)].

In the airplane these requirements are met in two ways. The first is the simulated power loss. In this scenario the engine fails and the airplane become a heavy glider. This scenario can be played out in an airport's traffic pattern, in which case the pilot attempts to land on the actual runway with no power, or it can be simulated away from an airport, which requires the pilot to select a suitable field for landing. In either case, the pilot does have some time as the airplane glides down. The private pilot PTS requires this demonstration. During the test the examiner may simulate a power loss and then determines if the applicant

1 Exhibits knowledge of the elements related to emergency approach and landing procedures
2 Establishes and maintains the recommended best-glide attitude, configuration, and airspeed, within ±10 knots
3 Selects a suitable emergency landing area within the gliding distance of the airplane
4 Plans and follows a flight pattern to the selected landing area considering wind, altitude, terrain, and obstructions (WATO)
5 Attempts to determine the reason for the malfunction and makes correction, if possible
6 Maintains positive control of the airplane at all times
7 Follows the appropriate emergency checklist

The second scenario is even more filled with hazard. Fire in flight or some other emergency requires the people in the airplane to get out as soon as possible. During the power-loss scenario, the pilot has the luxury of time. Not much time, but most light airplanes will afford several minutes of gliding time from a cruise altitude. But when the airplane is on fire, there would be no time to allow a best glide. If the best glide will get the airplane to the ground in 3 minutes, but the airplane will be completely engulfed in flames in 2 minutes, you don't want best glide. With flames on your toes, you want to aim the airplane at the ground and get down as fast as the airplane will fall. Most airplane manufacturers will include in the *Pilot's Operating Handbook* an emergency descent procedure, but you might test out various speeds and configurations yourself. I discovered that in the complex airplane that I teach commercial pilots and flight instructors I can get a much faster rate of descent by putting the landing gear down, the propeller full forward, pushing the nose over to the yellow arc, and leaving the flaps up than with a slower speed and the flaps down. You might have to experiment. But I know that I can get down fast, get on the ground, and run.

The private pilot PTS has a separate requirement for this type of urgent emergency. The examiner will test the applicant on emergency descents by observing the applicant

1 Exhibiting knowledge of the elements related to an emergency descent
2 Recognizing the urgency of an emergency descent
3 Establishing the recommended emergency descent configuration and airspeed and maintaining that airspeed within ±5 knots
4 Demonstrating orientation, division of attention, and proper planning
5 Following the appropriate emergency checklist

Other emergency scenarios are not so easy to simulate in flight. For these a "table-top" approach is the next best thing. Play the what-if game. Ask yourself, What if this happened? Read some of the accident reports. Place yourself in a situation where troubleshooting is required. I hope you never face a true emergency, but wouldn't it be better to face an emergency that you had already thought through?

Scenario: What if the throttle cable broke or worked loose? A pilot is flying home after a short cross-country flight. She enters the traffic pattern and begins the prelanding checklist. When she pulls the throttle back to reduce power, the power does not reduce. She moves the throttle again only to discover that the rpms are remaining constant at a cruise power setting. The engine is operating on its own with no apparent control from the throttle. What do you do now? Would moving the mixture control to cutoff and back to rich reduce altitude? Would you eventually pull the mixture all the way to cutoff position and leave it there for landing?

Scenario: What if the engine started to run rough? A pilot levels off in a light airplane at 10,500 feet MSL. The pilot has never been this high in this airplane before, but this trip takes him across a mountain ridge. The airplane seems more sluggish and slower than ever before. Then it happens. The engine misses and then repeatedly runs rough. What do you do now? Has the mixture been properly leaned for this altitude? Is the primer in and locked? Would you try applying carburetor heat?

Scenario: What if the door popped open immediately after takeoff? A pilot is leaving on a long-awaited vacation flight. He loads his family and all their luggage into the airplane and away they go. They must wait in line behind several other airplanes before it is their turn for takeoff, but finally they are number 1. Runway lined up, full power, airspeed alive. The airplane gets light, then breaks ground. Bang! A sudden loud rush of wind, flapping papers, and kids yelling. The airplane yaws to one side. At first the sounds catch the pilot off guard and it takes a few seconds to realize that the door is open. What do you do now? Do you continue climbout? Is there enough runway ahead to land? Can you get the door shut while in flight? What effect on airplane control will an open door have?

These are just a few examples. You could invent many more. Together we may not be able to create every possible emergency situation, but the more we think about, the better prepared we will be. The private pilot PTS also points a few out. On the checkride the examiner will determine if the applicant

1 Exhibits knowledge of the elements related to system and equipment malfunctions appropriate to the airplane provided for the flight test
2 Analyzes the situation and takes the appropriate action for simulated emergencies, such as
   a Partial or complete power loss
   b Engine roughness or overheat
   c Carburetor or induction icing
   d Loss of oil pressure
   e Fuel starvation

    f Electrical systems malfunction
    g Flight instrument malfunction
    h Landing gear or flap malfunction
    i Inoperative trim
    j Inadvertent door or window opening
    k Structural icing
    l Smoke or fire in the engine compartment
    m Any other emergency appropriate to the airplane provided for
        the flight test
3 Follows the appropriate emergency checklist

Emergencies really put pilot decision making to the test. The only good thing you can say about emergencies is that they readily lend themselves to scenario-based training. Emergency scenarios are great opportunities to practice decision making. Again, I hope you never are faced with an actual emergency, but if something ever does happen, you will be much better able to deal with the problem if you have played what-if with it already.

Checkride preparation

The practical test is the first time a student pilot will truly be pilot in command. Student pilots may log PIC time now, but they are still under the watchful eye of their flight instructor, and the instructor has veto power over the students' decisions. But during the practical test student pilots must act as pilot in command and their ability to assume that role is being graded. Every student, every pilot, will get nervous prior to a checkride. But you must realize that dealing with the anticipation and nervousness is part of the test. On occasion students who have not done particularly well on a stage check will tell me that they would have done better if they hadn't been so nervous. This statement is always a little scary to me. Are they saying that they are safe, proficient pilots unless they get nervous? It would seem to me that if something were to happen during the flight to make a pilot nervous, they would want to be at their best, not worst. If nervousness robs them of the ability to conduct the flight safely, they should never go up in the first place. I don't think any examiner could ever scare someone more than a rough-running engine could. Remember that examiners want to see if you handle a little stress on the practical test. Being nervous for the test plays right into their hands. Understand that dealing with the stress of the test and still performing safely is part of the test of being pilot in command.

All the scenarios of this chapter were derived for one purpose: to combine a maneuver with a situation in which decisions are needed. We cannot expect pilots to fly a maneuver without practice, and likewise we cannot expect pilots to make in-flight decisions without practice. Pilots must get to the point where they see decision making as just as big a part of the flight as takeoff. Students and instructors should borrow a page from the airline book and create "event sets" that apply to flight training topics. The event sets themselves come from everyday flying experiences.

10

Decision Training
for Instrument Pilots

The physical action of manipulating the airplane controls is a very small part of being a proficient instrument pilot. It is the thinking game and the decisions that are made that turns an "approach pilot" into an instrument pilot. This chapter is devoted to the ways in which students of instrument flight and current instrument-rated pilots can train to become instrument decision makers. As with the previous chapter, the examples that are discussed here are not the final word or the best way to do it. They are examples that should create some thought, which will lead to even better examples that pilots and instructors will invent.

This chapter also combines the findings of the work with general aviation pilots with the techniques of naturalistic decision making. Figure 10-1 illustrates the natural building blocks of instrument training. Each block represents a set of skills and applied knowledge. Each block depends on the one below it to provide a foundation. The skills and applied knowledge from the block below make it possible to move to the block above. There are no flight-hour minimum requirements within each block. This is an advanced qualification program (AQP) type of approach, where progression is based on acquiring skills and understanding topics, not on hours the engine has been operating. Running through each block places the pilot into decision-making situations. A pilot is not ready to take the next step until he or she can consistently make quality decisions that are presented within the block. In the end, pilots who are competent decision makers emerge.

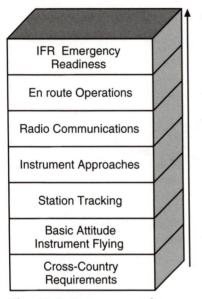

Fig. 10-1 *Instrument pilot building blocks.*

Cross-country requirements

Before pilots become eligible for an instrument rating, the regulations require that they practice their cross-country navigation skills. Specifically the regulation is FAR 61.65(d):

> *A person who applies for an instrument rating must have logged the following:*
>
> **1** *At least 50 hours of cross-country flight time as pilot in command, of which at least 10 hours must be in airplanes for an instrument airplane rating.*

This means that the instrument applicant must first do a lot of VFR flying. The instrument rating no longer has a stated minimum number of flight hours (it was 125 hours). But by the time a person completes the private pilot certificate with a minimum of 40 hours, flies 50 hours of cross-country time, and then the 40-hours minimum of instrument training, the applicant is still pushing 125 total. (Note: 40 + 50 + 40 = 130, but some of this time may overlap to produce a smaller total.)

What exactly qualifies as a cross-country for the purpose of meeting this 50-hour minimum? Recently revised regulation definitions have made this more clear. FAR 61.1(b)(3) says

Cross-country time means

ii *For the purpose of meeting the aeronautical experience requirements for a private pilot certificate, a commercial pilot certificate, or an instrument rating, time acquired during a flight [that qualifies as cross-country time must be]*

A *Conducted in an appropriate aircraft;*

B *That includes a point of landing that was at least a straight-line distance of more that 50 nautical miles from the original point of departure; and*

C *That involves the use of dead reckoning, pilotage, electronic navigation aids, radio aids, or other navigation systems to navigate to the landing point.*

So a private pilot who wishes to seek an instrument rating needs to make flights that are greater than 50 nautical miles. Anything less will not count toward the 50 hours. It says that the 50 miles must be a "straight-line distance." So you cannot fly 50 miles in a circle, and this also means that the total distance of the trip will be more than 100 nautical miles: +50 out and +50 back. It also is clear that the pilot must make a landing at a point that is more than 50 nautical miles away. You can't just fly past 50 miles and turn around and come back.

The only exception to the "must land" rule is for pilots working toward the airline transport certificate. FAR 61.1(b)(3)(iv)(B) says cross-country time is "at least a straight-line distance of more than 50 nautical miles from the original point of departure" but does not mention a landing. This means that even as a private pilot you should log as cross-country time any flight that takes you more than 50 miles out even if you do not land, because this can later be used to meet the airline transport pilot (ATP) requirement. If you do this, however, you should make a note of it, or, better yet, keep a separate logbook column for ATP cross-country. Several of us fought for this regulation change for a long time. I have employed pilots who took photographs of crops from the air. They flew hundreds of miles a day, but they always took off and landed back at the original airport. That job [before global positioning system (GPS)] required very skilled navigation, but it was not considered then as cross-country because they did not land while outside 50 miles. Also, instrument instructors will fly 50 miles, shoot an instrument approach that ends in a missed approach, and return home. Before, even though they may have been within 200 feet of the runway, it was not considered cross-country because they did not land. Many CFIIs made "break out of the clouds and touch-and-go" scenarios to remedy that problem. Today there is no problem and any cross-country time of this nature should be logged. The ATP cross-country requirement is 500 hours, and 500 hours of cross-country time can be hard to come by within 1500 total miles, especially if you are a flight instructor.

One last comment in this area. The ATP certificate is now easier to obtain. You can earn the ATP with only a third-class medical, and it can be conducted in a single-engine airplane. Essentially the check-ride is an airplane systems and instrument flight test. Instead of yet another biannual flight review, think about adding on the ATP this year. You will need to take the knowledge (written) test, but you could conceivably take the checkride in a Cessna 152. You will end up with a single-engine ATP, an instrument proficiency check, another 2 years of pilot in command, a possible break in insurance rates, and a great deal of respect.

Can cross-county flights be linked together? Yes, but each leg must be more than 50 nautical miles. If you wanted to take an extended trip and use that time toward the 50-mile requirement, you could do it, but you must measure out the legs. If you have one leg that is 100 miles, a 65-mile leg, and finally a 45-mile leg, that last leg would not count. The best way is to spread out the legs more evenly by selecting airports along the way so that each is more than 50 nautical miles from the last one.

Can the cross-country time that was flown by a student pilot while working toward the private pilot certificate be used toward the 50-hour instrument rating requirement? Yes. FAR 61.89(a) says "A student pilot may not act as *pilot in command* of an aircraft…" and lists eight stipulations. It is assumed then that if student pilots meet all eight stipulations, they can act as PIC and log PIC time. The private pilot certificate requires 5 hours of solo/PIC cross-country time. Those 5 hours can count toward the 50.

A private pilot who is working toward qualifying for the instrument rating will have 5 hours already and then should shoot for about 42 more. The 5 student pilot cross-country hours plus 42 more takes you to 47. The instrument training itself also has a required cross-country that will be at least 3 hours in duration. Put it all together and the pilot efficiently meets the requirement. I strongly recommend that these 47 VFR cross-country hours be accomplished before any of the instrument training begins. If you try to combine them, it is inevitable that the IFR work will get ahead of the VFR work. The day will come when you are ready for the instrument rating practical test but have to wait for good weather to complete a VFR cross-country. The delay will cost money when you must refly instrument flights to maintain checkride proficiency.

The regulation says 50 nautical miles but never stipulates where the pilot must go. I once worked with a private pilot toward the instrument rating who had never spoken to anyone on the radio except unicom. He had a very hard time with IFR work because radio communications were a huge hurdle for him. He had correctly met the cross-country requirements, but every airport he ever went to was an uncontrolled airport. Do not fall into this trap. Use the 50 hours of cross-country requirement to improve your radio skills and comfort level in congested airspace. Students with whom I work have a cross-country "pyramid" that I require them to climb. The 50 hours starts out with flights to uncontrolled airports, but with each successive flight the bar is raised. Next I have them fly to an airport that still has a "walk-in" flight service station. These are becoming hard to find, but I still have two in my region. The students must walk-in and take an FSS tour and then bring me back something from their tour (a weather chart usually). Next they fly to several Class C airports so that radio communications become normal and natural. Of course, sometimes I get a call from a controller who wants to discuss a particular pilot's technique, but I feel it's better to get out there and mix it up than to never go at all. Then comes VFR night cross-countries, followed by a flight across the Smoky Mountains. Before the target of 47 cross-country hours is reached, they will fly into and out of a Class B airport and take an extended flight that will also meet commercial pilot requirements (see Chap. 11). When all this is done, they are much more prepared to step into the IFR world than if they had spent the 50 hours flying back and forth between two familiar small airports.

In addition, having the pilots widen their experience meets the goals of real-world training, because they become immersed in the real world. I cannot possibly plan in advance all the experiences pilots will be exposed to during those 50 hours. They must go out and be faced with real decisions. Routinely being placed in a position where a decision is warranted is the best way to become comfortable with making decisions.

Basic attitude instrument flying

Following the completion of the 50 hours of VFR cross-country time, a pilot seeking an instrument rating begins the instrument training in earnest, with basic attitude instrument flying. Pilots who are already instrument rated also work with basic attitudes as they maintain proficiency over the years. Basic attitude instrument flying is not a favorite among pilots and students. Repetitious turns, climbs, and descents under an IFR simulation hood can become fatiguing. The maneuvers seem pointless because there seems to be no reason for them or destination, but being able to fly the airplane well with reference to only the instruments is vital. Look at the performance of the volunteer pilots in the simulator (in the appendix). The pilots who had the most problems were the pilots who had problems with attitude instrument flying. They could either fly the airplane or think, but not both. The airplane alone was taking all their attention, so they had no leftover brain space to think out problems. Figure 10-2 illustrates three pilots. The first pilot's mind is at rest. This is what we look like when we are at home watching a ball game with only passing interest. The second pilot is flying an airplane on instruments. Because of a lack of experience or recent practice this pilot has been almost completely saturated by the task of attitude instrument flying. The second pilot has very little leftover brain space. If a situation develops that requires this pilot to think out a problem, there will be trouble. Even a problem with a simple solution will be impossible as long as the pilot is already saturated. The third pilot is also flying on instruments, but this pilot has reached the ability level where attitude instrument flying does not saturate the brain. The airplane is being flown by a "mental autopilot." This leaves room to plan, anticipate, prioritize, deal with emergencies, and be pilot in command. All of us when flying want to look like the third pilot, but we all start out looking like the second one.

A creative instructor or pilot can place some real-world circumstances even into basic attitude instrument flying.

Fig. 10-2*a* *The brain at rest with plenty of available "brain space."*

Fig. 10-2*b* *The brain when almost completely saturated. There is little or no "brain space" left over.*

Fig. 10-2*c* *The brain on "mental autopilot with brain space to spare.*

Scenario: You are flying on instruments (probably under the IFR simulation hood practicing attitude instrument flight) and tracking inbound to a VOR station. You are at 5000 feet and a controller tells you to cross the 20 DME position at 3000 feet. You are currently passing the 30 DME position. Can you calculate the proper rate of descent? You can if you have an estimate of the ground speed or ask the controller to give you a ground speed readout. Use 120 knots ground speed for this example. How many miles are you traveling every minute? 60 knots is equal to 1 mile every minute, so 120 knots would be 2 miles every minute. The distance between the 30 DME position and the 20 DME position is 10 miles. Moving 2 miles each minute we will travel 10 miles in 5 minutes. That means that we have 5 minutes to descend from 5000 feet down to 3000 feet, which is a loss of 2000 feet. To lose 2000 feet in 5 minutes would require at least a 400-foot-per-minute rate of descent (2000/5 = 400).

The math in this problem is not hard, and while sitting on the ground it is easy to reason out the problem and come up with the answer. But can you do it while simultaneously flying the airplane? This type of descent instruction is very common from ATC. Every instrument pilot should be able to deal with this request, and it is a great example of attitude instrument flying in action. The pilot will need to hold heading and continue tracking the VOR radial with whatever wind correction might be required. The airplane's pitch must be adjusted to achieve a descent rate. The power might also need adjustment for a cruise descent. All this is going on with reference to instruments alone, and all while the pilot is simultaneously calculating the proper descent. This is a great exercise that combines instrument skills and real-world air traffic control instructions. It also tests how much free brain space is available.

Station tracking

The next basic skills that must be solid with instrument pilots is the ability to make intercepts and track courses; after all, every instrument approach that a pilot might fly in the future is a straight course track. We learned in geometry class that the shortest distance between two points is a straight line, and the beauty of airplane travel is that we travel faster because we can travel on a straight line. Tracking a station is flying the straight line.

Scenario 1: You have been asked to take aerial photos of a county line so that your friend, the county commissioner, can make a rezoning decision. The county line is not actually marked and cannot be seen from the air. The only way to know for sure that the photos are of the correct line is to fly a perfectly straight course over the line. A nondirectional beacon (NDB) happens to sit exactly on the county line (imagine that). Figure 10-3 illustrates how to take the correct photos.

If the county line extends east and west of the NDB, you must fly the airplane exactly on a bearing of 090° if approaching from the west. The trick will be to get over the county line in the first place. Figure 10-3*a* shows the airplane approaching the county line on a heading of 135°. How will you know when you are over the line and should turn and fly down the line toward the NDB? The inbound course is 090°. The intercept heading selected is 135°. The difference is 45°, which is the intercept angle. You will be over the county line when the intercept angle and the automatic direction finder (ADF) needle

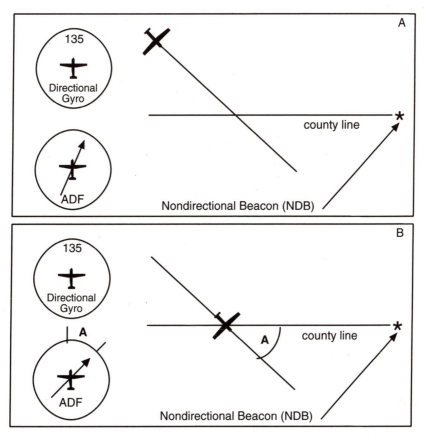

Fig. 10-3 *Course intercept using the ADF.*

deflection angle are equal. Figure 10-3*b* shows that the airplane has flown into position on the intercept heading of 135°. The intercept angle is shown here with the letter A. Meanwhile on the ADF the deflection needle has "fallen" to a position 45° to the right of the airplane's nose. This angle is also shown here with the letter A. The intercept angle is 45°, and when the ADF needle also becomes 45°, the angles are equal and the airplane is over the county line. You turn to a new heading of 090° and the ADF needle will point the station out on the nose (Fig. 10-4*a*). Start taking pictures now.

As long as the heading is 090° while the ADF needle is on the nose, you are right on target, as shown in Fig. 10-4*a*. But if a north wind were present, could the airplane still fly over the line? If you hold a heading of 090° with a wind from the left (north), the airplane will eventually drift "downstream" as shown in Fig. 10-4*b*. The ADF needle no longer shows the station just ahead, but instead the station is off to the left of the nose. The airplane is no longer over the county line and you are not taking the correct pictures. The entire county might be incorrectly rezoned because of this error. What should have happened? You should have included a wind correction angle to the original track. Figure 10-4*c* illustrates the airplane flying with a heading slightly into the wind at 075°. This heading change into the wind should compensate for wind drift and keep the airplane over the line. After the heading change the NDB station is no longer right in front of the airplane. Now the station is slightly to the right of the nose. As long as the wind correction angle and the ADF needle deflection angle are equal, you will stay right over the line all the way to the station. You know now that you will be bringing back the correct photos. This is called tracking. The airplane will fly a straight, not a curved, course to the station.

Tracking to a VOR station is slightly easier because all the angles that must be calculated when intercepting and wind correcting are done for you when the course deviation indicator (CDI) needle centers. But this ability to track to a station is vital to instrument flight. Go out and track a line (it does not have to be a county line); this will also test your ability to fly a level altitude, hold a precise heading, and think while doing it.

Instrument approaches

Pilots working toward their instrument ratings start to feel they are making progress when they begin flying instrument approaches. Pilots

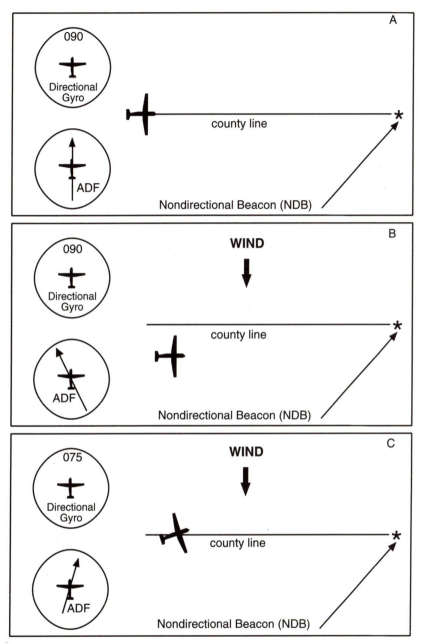

Fig. 10-4 *Course tracking using the ADF.*

who are already instrument rated must use instrument approaches to maintain instrument currency. FAR 61.57(c) says

> *No person may act as pilot in command under IFR or in weather conditions less than the minimums prescribed for VFR, unless within the preceding 6 calendar months, that person has:*
>
> **1** *For the purpose of obtaining instrument experience in an aircraft (other than a glider), performed and logged under actual or simulated instrument conditions, either in flight in the appropriate category of aircraft or in a flight simulator or flight training device that is representative of the aircraft category for the instrument privileges sought*
>
> > **i** *At least six instrument approaches;*
> >
> > **ii** *Holding procedures; and*
> >
> > **iii** *Intercepting and tracking courses through the use of navigation systems.*

In short this means that instrument pilots must fly six approaches and complete a holding pattern somewhere every 6 calendar months to go IFR. So instrument approaches are the emphasis in training and for proficiency. It is ironic that in real-world IFR flight the percentage of the flight spent flying approaches is small. It might be a 2-hour flight, only 10 minutes of which is on the approach. But in training and in currency flying very little time is spent doing anything else but approaches. It is another example of the false sense that non-real-world practice produces.

But instrument approaches must be learned and practiced, and so it is a necessary evil to fly them repetitiously. In the beginning of instrument approach training you should fly a few approaches while purposely looking out the window at where you are going. Inbound, on a VOR approach for instance, you should be able to see the VOR station up ahead and picture the line that runs from the VOR to the runway. This will help you visualize what it should look like later when you can't see it.

Before you begin approaches training, make some ground rules and start with a set of understandings. First, if repeated instrument approaches are made, the reason for doing them that way is to save time and money. You will never see a corporate jet filled with busy executives making repeated instrument approaches in clear air. Repeated approaches are to facilitate cost savings in flight training but can produce a negative carryover (Chap. 7).

Second, understand what role the flight instructor will be playing during the flight. Sometimes the flight instructor is only the teacher, but sometimes again to facilitate an IFR simulation, the instructor also acts as an air traffic controller. In some circumstances the instructor might also act as a required crew member such as a first officer. When you take instrument instruction, either for the instrument rating or for refresher training, talk about the role that will be played when approaches are being flown.

Third, if you must use a view-limiting device (IFR hood, goggles, etc.) to simulate flight in the clouds, treat it like the clouds. Agree before takeoff that any time you are under the hood you will do everything just as if you were in the clouds. Also agree that you will continue to remain in the clouds until the instructor lets you know otherwise. Would you ever switch to the VFR transponder code of 1200 while flying in the clouds? Of course not, so don't do it when flying in simulated clouds under the hood. Only do in simulation what you would do in actual instrument conditions. This will do a better job of creating the real-world environment.

Fourth, don't turn instrument approaches into another set of maneuvers. Use a mission-oriented method.

Scenario: An instrument student and an instructor meet for their regular Saturday morning flight lesson. The agenda for this lesson is to fly a VOR/DME approach. The nearest VOR/DME approach is at an airport that is 20 miles away, so the two plan to make the trip up and back as part of the lesson. They file an IFR flight plan using the airport with the VOR/DME approach as the destination and their home airport as the alternate. The instructor tells the student that by coincidence the airport with the VOR/DME approach also is holding an airplane part that the maintenance department would like picked up. The instructor asks the student if it would be okay to make a full-stop landing at that airport and retrieve the part. In fact there is no part to pick up; the story is just the instructor's setup. The student and instructor take off and soon the student goes under the IFR hood. The instructor and student had previously agreed to honor the simulation of the hood and that the student was not to assume that the airplane was out of the clouds unless the instructor specifically said that they were. They fly the VOR/DME approach. When the airplane arrives at the MDA, the student levels off, but the instructor says nothing; they are still in the clouds. When the airplane reaches the missed-approach point (MAP), the instructor is still silent; they are still in the clouds. What do you think the student will do?

The student will be torn. He thinks that they *must* land because, after all, they must get the airplane part. On the other hand, they *must* make a missed approach if the runway is not in sight at the MAP. If the student truly believes the simulation, he will make the missed approach. If the student still is in the training environment, he will take the hood off and say, "Hey you forgot to make the clouds disappear!" You see, in actual IFR flight, no matter how badly you sometimes would like to, you cannot make the clouds disappear. If the student does take the hood off when he should not and blows the scenario, land and have a talk. If the student makes the missed approach, the scenario continues.

The instructor, acting as a controller, tells the student to report entering the holding pattern. The holding pattern referred to will be the one in the missed-approach instructions. Now the student must remember what the instructions were and how to get into the hold. Once in the hold, the student reports the entry, and then the instructor, once again acting as a controller, gives a very bleak weather report for the airport with the VOR/DME approach. Should the flight proceed to the alternate? The alternate airport was the home airport, the starting point of the lesson. The decision to go to the alternate is made even though it will be a tough decision—the maintenance technician will be disappointed that you did not come home with the part. But making the decision to disappoint someone is a decision that pilots make all the time. The instructor has fashioned the scenario in such a way that there are actual consequences to the pilot's decisions. Ultimately, the pair fly an approach back to the home airport. This time at the MAP the instructor tells the student to remove the hood. The runway then is actually in sight and the landing is made.

That mission-based scenario contained the VOR/DME approach, a missed-approach decision, a holding pattern, an alternate airport decision, en route procedures both over to and back from the VOR/DME approach, and then an approach back home. Using the scenario would not have made that flight lesson cost any more than it would have otherwise, but the student would have gotten much more for the money.

Scenario: After an instrument approach is missed, a pilot is told by the controller to "execute the published missed-approach procedure and report entering the missed-approach holding pattern." The pilot complies. The hold entry is made and the pilot reports in, "Approach 56 Delta has entered the hold over the outer marker." The controller says, "Roger, 56 Delta, maintain 3000."

What should the pilot do? The controller has left something extremely vital out of the missed-approach and holding instructions. There has been no expect further clearance (EFC) time given. The EFC is critical. It spells out the time in which the airplane can leave the holding fix and proceed on to the destination or begin an instrument approach. What if a pilot accepted the hold instructions as this controller has stated them and while holding in the clouds, the two-way radios failed? Without an EFC the pilot does not know what to do. The pilot cannot hold forever. Eventually the airplane will run out of fuel. The pilot cannot just leave the hold at anytime, because she will then plunge into the very air traffic problem that caused her to hold in the first place. The pilot is stuck with no safety net. But, when an EFC is issued, and two-way radios fail in the hold, the pilot will remain in the hold until the EFC time runs out and then fly to the destination for an instrument approach and, it is hoped, a problem-solving landing. The controller should say, "Hold as published, expect further clearance at 25, time now is 10." The 25 means 25 minutes past the hour and the time now is 10 is an invitation to synchronize your watch with the controller's time. If the controller does not offer an EFC, the pilot should ask for one: "Hold as published, roger, and can I get an EFC on that?" Although I have had controllers balk on occasion, they usually will bring the EFC right back.

The EFC offers some problems for flight training. If a pilot is holding but is actually in VFR weather conditions when two-way radios fail, there is actually no problem. The pilot can leave the hold and fly visually to the next practical airport. VFR conditions render the EFC somewhat moot, and controllers know this. The problem comes up when a pilot and instrument instructor are flying in VFR conditions but simulating IFR conditions. During the simulation we would prefer controllers to play along and handle the situation like it was really IFR, which would include issuing EFC times for holding patterns. But, controllers do not always play along. They know when it's really VFR, and even though you may be on an IFR clearance, they will handle traffic very VFR-like. This means that often when

IFR holds are practiced in VFR conditions, EFCs may not be forth-coming from the controller. This is another example of a problem that lies within the training environment that does not exist in the real environment. Instrument instructors should use good judgment here. To protect the IFR simulation, it would be best to always get an EFC when holding but do not badger an overloaded controller with a request that seems unnecessary in VFR conditions. If you must hold in VFR conditions and the controller does not follow the simulation with an EFC, make sure to explain this problem. And under no circumstances should you ever accept a hold in instrument meteorological conditions without an EFC.

Scenario: An instrument-rated pilot needs an instrument proficiency check from an instrument instructor to regain IFR currency. This pilot has flown from the same airport for over 10 years. The airport has only one approach: an NDB, and the pilot must have flown that approach what seems like 1000 times in those 10 years. The pilot asks an instrument instructor to fly with her for the purpose of meeting the proficiency check requirements. The CFII agrees, and a date is set. When the day comes the pilot prepares her co-owned airplane for flight and meets the instructor. Soon they are on the runway and away they go. Five hundred feet into the air the instructor asks to pilot to put on the IFR hood. The pilot puts on the hood and then pulls from her flight bag a copy of that airport's NDB approach, which she has had laminated. The pilot does not seem to have any other charts. "Okay," the pilot says, "I'll turn to the station and get started." The instructor returns with, "Let's do this a little different this time. Let's say that instead of beginning this approach, that we have just completed it and the clouds were too low to land. We are now on the missed approach and faced with shooting a different approach to a different airport."

The pilot, not playing the game at all, says, "No, I can only be up about 1 hour; we can't go anywhere else." The instructor pulls from his bag an approach chart book. The instructor explains that they do not have to go anywhere else to shoot a different approach. They are going to use their hometown NDB, but shoot an NDB approach using a chart from another airport. Everything about the approach will be the same except for the frequency of the NDB. The instructor had previously selected an NDB approach that had safe altitudes to fly over the home airport and that would not interfere with the home airport's traffic.

The first problem the pilot encounters is that the approach the instructor wants her to shoot is in a chart book. Will the pilot be able to fly the airplane and find the chart in the book? Remember, the pilot had eliminated this step by preparing only one chart, but in real-world operations you never know which approach you might be called upon to shoot. Airplane control suffers, but the pilot eventually locates the requested approach chart. The next problem the pilot detects is that at her hometown airport the NDB is located on the field. The approach that the instructor has chosen shows the NDB off the field and the approach requires timing from the NDB to the airport. How will the pilot do on this unexpected approach? Will the pilot and the instructor remain friends after the flight?

This is a real-world setup: an unexpected approach. Pilots and instructors alike can practice any VOR approach over any VOR and any NDB approach over any NDB. If you use this technique, exercise caution. You would not want to select an approach that has an MDA that is too low to the actual ground or that crossed a traffic pattern. I live in an area that has relatively low ground elevations, so I use approach charts from areas of the country with high elevations. Choose any NDB approach from the state of Colorado and the altitudes on that approach will be much higher than any NDB approach in Tennessee. When I shoot the NDB approach using the chart for Pueblo, Colorado, I will be way up high and out of the way of the traffic pattern or anyone on the actual NDB approach below. I will have students flying approaches from across the country. In this way they will not become complacent with the same old approach over and over again. This simulation must be done in VFR conditions using an IFR hood, however. You can't actually ask a controller who works the airspace over middle Tennessee to give you vectors for an approach that is in Pueblo, Colorado.

Radio communications

In every instrument pilot's life there is one flight lesson that he or she will never forget. This lesson is predictable and inevitable. It is the first time that the instrument instructor lets the student handle all the radio work. This is usually a memorable lesson because the wheels seem to fall off. When an instrument student is working in the instrument approaches block (Figure 10-1), the instructor usually is either helping with or completely accomplishing the radio communications. The student hears the radio but is too busy flying the approach to talk and is very happy to let the instructor take care of that. But to move to the next block, the microphone must move to the pilot's side. The first time that the instructor steps back and lets the pilot try his or her hand at the whole thing, watch out.

There is another reason for this lesson to cause problems. Pilots in the instrument approaches block can get a little overconfident. They get to the point where they can fly the approaches very well and are wondering when the instructor will schedule a checkride. When this attitude meets up with "the whole thing," the pilot is in for a surprise. The most common thing said after the whole thing flight is, "I don't know what happened." The student honestly thought that he or she had become an instrument pilot because he or she could follow instructions on the approach chart. The skillful instructor will use this disaster to help students see exactly what they are up against and to see, maybe for the first time, the high level of challenge that is contained in the instrument rating.

Radio communications is much more than talking on the radio. Radio communications is where the pilot gets the information used in planning, organizing, and prioritizing the flight. A big part of situation awareness comes from simply listening on the proper ATC frequency.

Scenario: A current instrument pilot, flying a Cessna 172, N1234A, is being vectored for the ILS runway 2L approach. 34A is on an IFR clearance and is currently in the clouds. The radio communications over the ATC frequency goes like this:

"Nashville Approach 1234 Alpha checking in at 4000 with information x-ray."

> *"Roger 34 Alpha expect the ILS 2L at Nashville, descend and maintain 3000. Break. Delta 242 I need your best rate to 10,000, turn right on course"*

> *"On course for Delta 242. Looking for higher."*

> *"Nashville Approach Baron 6789 Bravo, 4000 with x-ray."*

> *"89 Bravo, expect the ILS for 2L, descend to 3000 turn right heading 270."*

> *"3000 and 270, 89 Bravo."*

> *"Nashville Approach, Lear 4567 Charlie off runway 2R, through 1500 for 4000, over."*

"Lear 67 Charlie, radar contact, climb to 12,000, turn right on course."

"Up to one two thousand on course, 67 Charlie."

"34 Alpha turn left heading 090."

"Left to 090 for 34 Alpha roger."

"Baron 89 Bravo turn to 330, descend to 2500"

Stop. Where is the Baron? Which airplane will fly the ILS first? Where is the Delta? Where is the Lear Jet? The only way you can know the answers to these vital questions is to listen and to understand. Many pilots just ignore or cannot keep up with the constant dance of the air traffic. They only seem to wake up when they hear their own airplane's number, and then they carry out instructions oblivious to what is taking place around them. Back to the scenario. The Delta and the Lear are no factor. Both took off going the other direction, flew quickly through 34A's altitude, and were gone. The Baron on the other hand is of interest, if not concern. Both 34A and 89B were told to expect the ILS 2L. 34A, the Cessna 172, was on the frequency and received these instructions before 89B, but 89B is a Baron and is therefore faster. At one point 89B is flying a heading of 270° while 34A is flying a heading of 090°, and both are going to the same approach. We are not concerned that they appear to be heading at each other, but from this information we know that 34A is west of the localizer while 89B is east of the localizer. Now it's a race. Who will shoot the approach first? The Baron's last instruction was to turn to 330° and descend to 2500. The Baron appears to be number 1 and 34A number 2. Why? Because 330° would be a logical intercept heading for the runway 2L (50° intercept) and 2500 is a lower altitude. Meanwhile, 34A has not yet been given an intercept heading. 34A is still flying 090°. The pilot of 34A should expect at least one more left turn to provide a better intercept angle, but the turn will not come until the Baron has passed. 34A is slower and still at 3000 feet, which is also an indication of who will be number 1. Did you get all the clues? Could you have understood all the clues while flying in the clouds, in the rain, in turbulence?

Radio communications is more than talking and knowing the proper phraseology. Radio communications is the door to situation awareness, and with awareness comes better decisions. Pilots must listen and then place all these airplanes on a mental radar screen so that they can be "seen" by the pilot in command.

En route operations

Private pilots working toward the instrument rating are required to complete an instrument dual cross-country flight. FAR 61.65 (d)(2)(iii) states:

> *For an instrument-airplane rating, instrument training on cross-country flight procedures specific to airplanes that includes at least one cross-country flight in an airplane that is performed under IFR, and consists of*
>
> **A** *A distance of at least 250 nautical miles along airways or ATC directed routing;*
>
> **B** *An instrument approach at each airport; and*
>
> **C** *Three different kinds of approaches with the use of navigation systems.*

Figure 10-5 illustrates this cross-country requirement. It must have a total distance of 250 nautical miles. It must have three different kinds of approaches. This does not specifically say three airports, but I always use three airports to get the job done.

I set up the whole lesson as a flight to just the first airport, using the second airport as an alternate. It takes most of a day to play this scenario out. We hope to have a day with actual IFR conditions to fly in. In the past, I have delayed this flight a day or two to specifically avoid good weather, opting for IFR weather instead. I select a nonprecision airport as our destination and an airport with an ILS as our alternate. This is exactly how it would be set up in the real world. I don't think anybody would choose as an alternate an airport that did not have an ILS. Would you want to fly to the destination airport, be surprised by a missed approach, and then fly with limited fuel to an alternate with just an NDB approach? Not me. To be used as an alternate there must be a weather observer on the field at the time. Most places that are large enough to have a weather observer are also large enough to have an ILS. The scenario will be that we cannot get below the clouds at the first destination airport and are forced into a decision to fly to the alternate. At the alternate, an ILS approach is flown that concludes with visual contact with the runway and a landing. I stop for fuel, but the scenario continues even on the ground. We are on the ground at the alternate, but we never got where we wanted to go—the first destination. We don't want to be stuck on the ground at

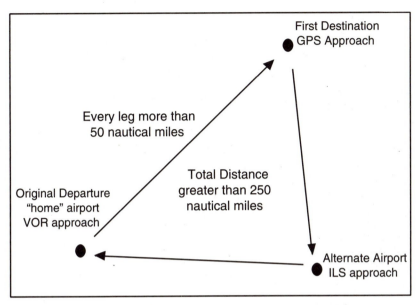

Fig. 10-5 *Diagram of "long" cross-country requirement for instrument rating.*

the alternate overnight, so our choices are to try for the first destination again or simply fly home. To help make this decision we check the weather. In the scenario the first destination weather is still too low, but the home airport weather is acceptable. We decide to fly home. The airport at home has yet another instrument approach, which we shoot to a landing below the clouds.

I am careful to plan the trip to exceed 250 nautical miles. I also plan so that every leg is greater than 50 nautical miles. That way, the student can use this flight toward the 50 hours needed to meet the instrument cross-country requirement. Properly planned, this flight will put the student over the 50-hour requirement.

The regulation for this lesson calls for an instrument approach at each airport and that three different kinds of approaches with the use of navigation systems be used. This allows for any combination of NDB, GPS, LOC (localizer), LDA (localizer type directional aid), SDF (simplified directional facility), VOR, VOR/DME, PAR (precision approach radar), and ILS approaches to be used. The regulation is also written so when new systems are developed, they are already included in the law.

That is what is required for training toward the instrument rating, but what about everyday work in the system? En route operations can be just about anything from departure to touchdown. Here are a few examples of situations that could be included in the original instrument training or in proficiency work with pilots who are already instrument rated.

Scenario: An instrument-rated pilot is en route to a small general aviation airport that has only a VOR approach. The flight is being conducted in instrument conditions on an IFR flight plan. The pilot wants to land at the airport, pick up a package, and fly back out. Knowing that this would be a quick turn-around, the pilot had originally filed two IFR flight plans: the one he is flying on now and another for the flight back out. The pilot gets within range of the airport and the controller says, "7146Q you are cleared for the approach, cancel IFR on this frequency or on the ground through Flight Service." Upon hearing this, the pilot switches to the small airport's advisory frequency and starts the descent. The pilot breaks out from under the clouds at the MDA, and while flying level to the MAP, the pilot sees the runway. Quickly the pilot switches back to the ATC frequency in hopes of canceling the IFR clearance while in the air. But the controller does not respond to the pilot's calls, and the airplane is now too low for direct communications. The pilot lands at the airport. Once out of the airplane the pilot calls 1-800-WX-BRIEF, gets an FSS briefer and cancels the IFR flight plan. The pilot picks up and loads the package into the airplane and is ready to go. The pilot remembers that on the way in, radio communications had been lost while still in IFR conditions. This little airport has no clearance delivery, ground control, control tower, or flight service station. How will the pilot ever get an IFR clearance to get out of there?

FAR 91.173 says

> *No person may operate an aircraft in controlled airspace under IFR unless that person has*
>
> *(a) Filed an IFR flight plan; and*
>
> *(b) Received an appropriate ATC clearance.*

In this scenario the pilot did file an outbound IFR flight plan but cannot get the clearance over the radio because radio reception is higher than the clouds. Another way to get the clearance would be over the telephone. The old-fashioned way is to call the FSS from a pay phone and ask them to get the clearance for you. The FSS briefer will put you on hold and call ATC. The controller will relay the clearance to you through the briefer and give a clearance "void time." You then must be off the ground before the time that the clearance is void. The void time is usually only about 10 minutes, so you must hurry into the air. Hurrying into the air is probably not safe, so old-fashioned void times can be frantic and hazardous. The new and much better void time method uses your portable cellular phone. If your cellular phone has service at the airport, you can complete the preflight inspections at normal pace. Then taxi out to the runway and complete pretakeoff checks carefully and at normal speed. Then call 1-800-WX-BRIEF for the FSS or direct to ATC if you know the number. You can get the clearance over the phone from the runup area and will need only a short void time. Is it approved to use a cellular telephone from an airplane? As of this writing it is only approved when the airplane is on the ground, and surprisingly this is not an FAA rule. FAR 91.21 says that electronic devices are not to be used in an aircraft, "while it is operating under IFR." This alone would mean that cellular phones could be used in flight while flying VFR, but Federal Communications Commission (FCC) regulation 22.911 prohibits cell phone use while in flight in any type of flight condition. However, receiving a void time clearance would take place on the ground, so can a cell phone be used in an aircraft while on the ground? There is no FAA regulation that speaks directly to this question, but a letter from FAA legal counsel Thomas C. Accardi to all FAA Flight Standards District Offices on February 4, 1992, made it clear at least to FAA personnel that cellular telephones operated in aircraft while on the ground were legal and in accordance with FAR 91.21(b)(5), which says that a device may be used when "the operator of the aircraft has determined [that the device] will not cause interference with the navigation or communication system of the aircraft on which it is to be used." So charge your cell phone and use it to receive the clearance.

What if the pilot who was trapped under the clouds at the uncontrolled airport could not communicate by telephone? The pay phone is out of order and the cellular phone is dead. To make the scenario even more interesting, the package that the pilot was picking up was actually an ice chest with a human heart inside. The flight is being made to deliver the heart for transplant and a life is at stake. What does the pilot do now?

Remember the exact wording of the IFR clearance rule. "No person may operate an aircraft *in controlled airspace* under IFR unless that person has (a) Filed an IFR flight plan; and (b) Received an appropriate ATC clearance." Could the pilot take off, climb into the clouds, and get high enough to achieve radio reception and accept the clearance without ever entering controlled airspace? It would depend on how high the Class G (uncontrolled) airspace extended and how close the nearest ATC frequency antenna was. In the scenario the pilot flew a VOR approach to land at this airport in the first place. That probably means that the airport is under a 700-foot AGL transition area. Seven hundred feet is not very high, but could the FSS be accessed through the same VOR that afforded the approach? If so, communications could be established as low as the approach's MDA. What about the surrounding terrain? This decision would call on all the pilot's resources. Flying in the clouds down low is certainly hazardous and should never be attempted unless it is clear to the pilot in command that it could be accomplished safely.

The "clearance in controlled airspace" rule comes into play with normal over-the-phone void times as well. Usually the clearance will start out like "ATC clears N1234A as filed, enter controlled airspace on a heading of 360°...." I guess they don't care what you do before climbing into controlled airspace as long as you enter going north.

Scenario: A pilot flying in the clouds and on an IFR clearance is approaching a busy Class C airport. The pilot has arrived during "the push" when many airplanes are converging on the same airport. The pace of the radio communications is rapid and tense. Airplane after airplane is being vectored and eventually sent down the funnel to the instrument approaches. The pilot then hears his airplane number called out, "5070R you fly heading 230, maintain 3000." This instruction is delivered with a clear and forceful tone. The pilot turns to 230° and holds at 3000 feet. The pilot then realizes that the heading of 230° accomplishes nothing. It is not an intercept heading for any approach. It is not a heading that even puts the pilot in a sequence for the approach. Several minutes pass and the pace of radio communications remains rapid and constant. The only difference is that the controller seems to be talking to everyone but the pilot in the story. What should the pilot do now?

Scenario: A pilot is flying a Piper Cherokee 180 in the clouds on an IFR clearance. The ride has been rough and the rain has been steady. The pilot notices that it is starting to get darker outside and the ride is more turbulent. Soon the rain sounds like bullets on the windshield and the airplane is being tossed about. What should the pilot do now?

Flying IFR without onboard radar is yet another challenge. The last thing a pilot wants to do is fly into an embedded thunderstorm, but how do you miss them if they are right on the route of flight and you cannot see ahead? First, tell the controller that you are without radar (control should know this already because you are in a Cherokee, but tell them anyway). Ask if they "paint" any precipitation echoes on their screen along your route of flight. They will tell you that their radar doesn't show precip very well, and this is true. The newest radars are actually round computer screens that present a composite from several radar antennae. The computer program removes most of the precip echoes so the airplanes can be seen better. The controller can adjust the computer so that slashed lines appear where light precip is present or the letter H will appear where heavy precip is present. Even these do not give very good information about the boundaries and contours of the precip, so controllers have helpful but incomplete information. Ask if any other aircraft have passed along your route. The controller might pass along a pilot report. If you hear a pilot passing a position that is on your route but still up ahead, you can solicit information from that pilot through the controller.

The best thing is to start working on the problem before you are facing the embedded thunderstorm. With time on your side, ask the controller if you can leave the frequency to contact FSS. You will usually be given a time limit to report back. Then switch and call the flight service through a VOR or a remote communications outlet or even direct if you are close enough. The FSS briefer has weather radar that can pinpoint the location and movement of the precip echoes. The weather radar does not show the location of airplanes. So the briefer cannot tell you where a cell is in relation to you, but if you know where you are already, you can judge your relative position to the storm. Formulate a diversion plan while talking with FSS, and when you go back to the ATC frequency, tell the controller your plan. When thunderstorms are around, you will be given just about any deviation you ask for. This is a great example of using crew resource management even when you are alone in the airplane. Make the controllers and FSS briefers members of your crew when they can provide essential information that you can't get on your own.

IFR emergency readiness

When the volunteer general aviation pilots participated in the project (see the appendix), they did not know what they were in for. In their first session in the simulator the alternator failed, and many were caught way off guard. In the second session, they must have known something was coming, but still many did not handle an oil loss and rough-running engine emergency very well. Most instrument pilots do well as long as nothing extraordinary happens. But it is the extraordinary circumstances that puts the test to any pilot in command.

Scenario: An instrument-rated private pilot flying a Cessna 172RG in the clouds notices that a thin bead of milky rime ice has formed along the wing's lead edge. The pilot looks back over her shoulder to see more ice on the horizontal stabilizer. Other than the pitot heat the Cessna 172RG has no ice protection equipment. The outside air temperature (OAT) gauge is reading just colder than the freezing mark. What action should the pilot take?

With no ice protection equipment, the only defense against ice accumulation is to get out of the icing conditions. Fortunately, icing layers in stratus clouds are usually thin. A climb or a descent should solve the problem. If terrain will allow it, a descent in a Cessna 172RG would probably be best if temperatures warmer than freezing exist at a lower altitude. If the terrain does not permit a descent, climb to a colder altitude or better yet climb into clear air. Tell the controller that you have picked up ice and make your request to change altitude. Make this request at the first sign of accumulation so that the controller can start working on the problem. If you wait too long, the climb or descent to safety could be delayed. When flying in the clouds, watch the OAT gauge as you climb. It may be necessary to fly the OAT rather than the altimeter, meaning the decision to level off in the climb depends on where the freezing level is, not where a cruising altitude is. If you climb through an icing layer to escape, the chances are high that you will accumulate ice on the way back down. Plan a steeper and faster descent than normal to cut down the exposure time. Ice is more hazardous on landing than at any other time. When ice gets on the airplane, the shape of airfoils and stabilizers is altered, so lift will be reduced. This lift reduction will become most evident when you slow down for the approach and landing. So if you have ice on the airplane during the approach, don't slow down. Try to maintain the cruise speed as long as possible.

Scenario: A pilot flying on instruments notices that the turn coordinator indicates a turn to the left and at the same time the attitude gyro indicates a turn to the right. The airplane cannot be turning both left and right, so what is happening? What should the pilot do?

It is true, you can't turn left and right simultaneously, so one of these two instruments is wrong. The trick is to determine which one is incorrect and start relying on the accurate one. The turn coordinator and attitude gyro are from different systems, so probably one system has failed or been impaired. First check the directional gyro (DG). The attitude gyro and the directional gyro both work on the vacuum system. If the DG is indicating a turn in the same direction as the attitude gyro, it would appear that they were working correctly and the fault must be in the turn coordinator. If the DG is not turning as the attitude gyro would indicate, it would appear that both have failed and the fault is in the vacuum system. This analysis needs to take place in the first second that suspicion arises because a pilot can become disoriented very quickly in this situation. Also, this situation is hard, if not impossible, to train for. Usually when instructors want their students to practice no-gyro flight and approaches, they cover the attitude and directional gyros with a piece of paper or a soap dish cover. When students see that these instruments are covered, they have no choice but to look at something else, like the turn coordinator and the magnetic compass. There is no "detection" being practiced. When the instruments are covered, the decision to look at other instruments is already made. In the real world, if the vacuum system fails, the instruments will not be magically covered from view. They will still be there, plain to see and be trusted. A real in-flight vacuum system failure is much more dangerous than anything that can be practiced because it is hard to practice detection. I always wanted to install a vacuum system relief valve that I could covertly open during an instrument flight lesson. The air flow would then bypass instead of turning the gyros, and the gyros would slowly fail. Then we could see true instrument detection or the lack thereof. Vacuum system failures do work well in the simulator. When pilots in the simulator are given a failed gyro system, they almost always eventually lose control of the airplane/simulator. It is that dangerous. The newest instrument rating practical test standards address this problem directly. Instrument-rating applicants are required to fly one nonprecision approach without the attitude and heading gyros. This still tests the student "after the fact." The failure is obvious when the instruments are covered, so the test is not about detection but about flying and dealing with reduced capability. The PTS says

Loss of Gyro Attitude and/or Heading Indicators

Objective: To determine that the applicant:

1. *Exhibits adequate knowledge of the elements relating to recognizing if attitude indicator and/or heading indicator is inaccurate or inoperative, and advises ATC or the examiner.*

2. *Advises ATC or the examiner anytime the aircraft is unable to comply with a clearance.*

3. *Demonstrates a nonprecision instrument approach without gyro attitude and heading indicators using the objectives of the nonprecision approach task.*

The solution to the detection problem is a cross-check. An instrument pilot can never let a situation continue where instruments disagree. You cannot be climbing while descending or turning both left and right. A constant vigilant instrument scan saves lives.

Scenario: An instrument-rated pilot is flying in IFR conditions. The pilot is level and en route to the destination with no problems. The flight continues through the clouds, but the ride is smooth and radio work is light. The pilot starts to notice that the air route traffic control center has been strangely quiet. The pilot calls the controller just to check in, but after several tries there is no answer. What should happen now?

Silence on the radio is a trap. I always want to hear a steady constant conversation going on over the frequency. Sometimes late at night, however, long time spans can go by without much chatter. Whenever I feel like I have not heard anything for a while, I will call the controller and ask some question, just to make sure we have each other. "Can you give me a ground speed readout?" or "What's the altimeter setting again?" or "Is my Mode C still at 8000?" Ninety-nine percent of the time ATC will call right back and there was never a problem. But if I have had a two-way radio communication failure, I want to know it sooner rather than later. If you suspect this problem, don't give up easily. First switch to another ATC frequency, maybe one that you had been talking to earlier in the flight. Try the other radio if you have two. Try any local frequencies that you might be flying over: a control tower, FSS, or even an AWOS or unicom. I was flying IFR once when the controller's transmitter failed. Of course, I first thought it was my radio that had failed, but I noticed that I could still hear and could communicate with all the other airplanes on the frequency. We all switched to other frequencies and talked to other controllers and there was no problem. Next, try to contact FSS through a VOR. If two-way communications have failed, you can still communicate through navigation radios that have voice capability. If all avenues to reestablish communications fail, fall back on the regulation that says you must go to your destination and hold if you are earlier than your estimated time of arrival or select and shoot an approach if you are on time or past the ETA. The savvy instrument pilot needs no greater hint here. You should never estimate your time of arrival on the flight plan correctly. If your time en route is 1 hour, you tell them 55 minutes. This way you will never be on time and you will never have to hold if two-way radios fail. It will be a bad enough day with the radios gone, so you don't want to hold on top of that. You always add about 5 minutes to en route estimates when filing VFR. This is to allow enough time to land, get out, and find a phone to call FSS and cancel, but IFR flight plans should subtract about 5 minutes.

A hand-held transceiver could come in handy in this situation, and many instrument pilots carry them for just such an occasion. I have noticed that their short antenna makes for short-range reception. However, I know an avionics shop that will sell you an exterior antenna hook up. This way you can unplug the failed radio from the antenna and hook the hand-held radio to the outside antenna for better range. What if you have not spent the money on a hand-held transceiver? Use your cellular phone. We discussed earlier how handy they have become on the ground. Well, they can become a life saver in the air when normal communications fail. Of course, using the portable phone in the air and in IFR is against FAA and FCC rules, but guess what—you don't care. FAR 91.3(b) says "In an in-flight emergency requiring immediate action, the pilot in command may deviate from *any rule* of this part to the extent required to meet that emergency." Two-way communications loss in the clouds is an emergency of the first order. What if there is an embedded thunderstorm ahead, but without onboard radar or two-way radios, you will not know any better than to plunge right into it. Call 1-800-WX-BRIEF from the air.

Cellular telephone reception towers are everywhere now, and there will be no signal blockage from your phone straight down to one of those towers. Chances are overwhelmingly favorable that the phone reception towers will be much closer than the nearest ATC reception tower. When the FSS answers the call, tell them your situation. They can relay information to ATC or better yet, they can give you the nearest ATC telephone number. Then you call ATC direct and say "Hey I'm the guy on your screen squawking 7600." The problem from then on is solved. Continue to the destination or get clearance for a closer approach.

There could be 100 more scenarios, and I'm sure you thought of a few while reading this. The point is that unless you ask the what-if questions and fly scenarios in your head in advance, you will be less prepared to meet the challenge if it ever actually happens to you. Remember, expert pilots are able to take emergencies in stride partly because they are familiar with the solutions in advance. They spend little time worrying about what to do about the problems. They apply the fix that matches the problem and go on. They know which fix to apply because they have traveled that road before. They might have never had that exact problem before, but they have played what-if before.

At every block through the instrument-rating process decisions are forced on the pilot. Maneuvers are flown, but decisions are emphasized. If an instrument pilot is faced with decisions on each and every lesson, making decisions will become a natural part of the process. A safe instrument pilot in command can fly the airplane, of course, but is also a natural and assertive decision maker.

11

Decision Training
for Commercial Pilots

The AQP idea can carry into general aviation's commercial pilot training as well. Figure 11-1 illustrates the major skill and knowledge areas that commercial pilots must master. The commercial pilot building blocks, like the others from previous chapters, start at the base and work up when standards of proficiency are met. No minimum flight hours are prescribed; when pilots can meet the standard, they move on. In the end, competent, decision assertive, commercial pilots are developed.

Cross-country requirements

Pilots don't much like to read the regulations. They are very hard to read with all their legalese. But not reading the regulations has cost many pilots some money when it comes to commercial pilot cross-country requirements. The fact is that most, if not all, of the commercial requirements can be accomplished while in the process of building the 50 hours of cross-country time needed for the instrument rating. But some would-be instrument pilots fly all 50 hours without reading the commercial requirements and never take a cross-country trip that kills two birds with one stone. When they do get to the commercial work and finally read the regulation, they discover that they must fly and pay for a flight that could have been previously incorporated. The regulations specifically are in part 91 under commercial pilot Aeronautical Experience. FAR 61.129 (a)(3) says commercial pilot applicants must have

> *20 hours of training…that includes at least*

> **(iii)** *One cross-country flight of at least 2 hours in a single-engine airplane in day VFR conditions, consisting of a total straight line distance of more than 100 nautical miles from the original point of departure; and*

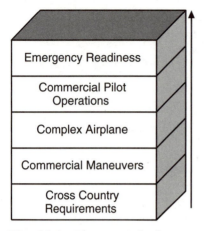

Fig. 11-1 *Commercial pilot building blocks.*

> **(iv)** *One cross-country flight of at least 2 hours in a single-engine airplane in night VFR conditions, consisting of a total straight line distance of more than 100 nautical miles from the original point of departure.*

This part of the regulations calls for these day and night cross-country flights to be "training" flights. FAR 61.1 defines "training" like this: "Training time means training received (i) in flight from an authorized instructor; (ii) on the ground from an authorized instructor; or (iii) in a flight simulator or flight training device from an authorized instructor." So these two flights must be conducted with a flight instructor and logged as dual received by the commercial applicant.

Then there are solo cross-country requirements. FAR 61.129(a)(4) says that commercial pilot applicants must have

> *10 hours of solo flight…that includes at least*

> **(i)** *One cross-country flight not less that 300 nautical miles total distance, with landings at a minimum of three points, one of which is a straight line distance of at least 250 nautical miles from the original departure point.*

It seems that whoever wrote this rule had a hard time with the math. If you fly out from the original departure point for 250 nautical miles and then eventually return home, you would have flown at least 500 miles (not 300). If after the first 250-mile leg the pilot elected not to return to the original departure point, the trip would still have to be 350 miles because there must be three legs and every leg must be at least 50 miles. The 300-mile total distance requirement is like false advertising. The rule is written into the FARs to maintain alignment with the International Civil Aeronautics Organization (ICAO) regulations. Annex 1 to the Convention on International Civil Aviation requires this commercial pilot cross-country, and having it as part of the FAR as well ensures that United States commercial pilot certificates are accepted internationally.

Figure 11-2 gives two versions of a flight that would meet this requirement and presents a decision to be made. I recommend that the long 250 nautical mile leg be conducted downwind. To save some money and/or to build some time, you should take this flight in something slow. But 250 miles against the wind in a small airplane is not comfortable, so let the wind plan your route. Figure 11-2*a* illustrates the recommended course with a wind from the west, and Fig. 11-2*b* shows the same course flown with a wind from the east. This method will divide up the headwind legs. Again, this flight can be accomplished prior to the instrument rating and count toward both the 50 hours that the instrument rating requires and the commercial pilot requirement.

The following is not a cross-country requirement per se but a new rule that requires night landings at a control tower airport. FAR 61.129 (a)(4) (ii) requires "5 hours in VFR night conditions with 10 takeoffs and 10 landings (with each involving a flight in the traffic pattern) at an airport with an operating control tower." The part that says that "involving flight in the traffic pattern" prevents pilots in Cessna 152s on a 10,000 foot runway from making four landings per pass. It would also be convenient to make some of these 10 landings to a full stop so that night currency is maintained, and some touch and go's for speed. The problem is that in the summertime, the VFR towers are closed after dark and the larger airport controllers do not know what a touch and go is. Apparently, the FAA did not brief the controllers on this regulation change before it became the law, so it is hard to get these all in during a single flight.

Decisions fill these cross-country and night requirements. Where should you fly? What route is best? What control towers are open at night? Where can fuel be purchased after normal business hours?

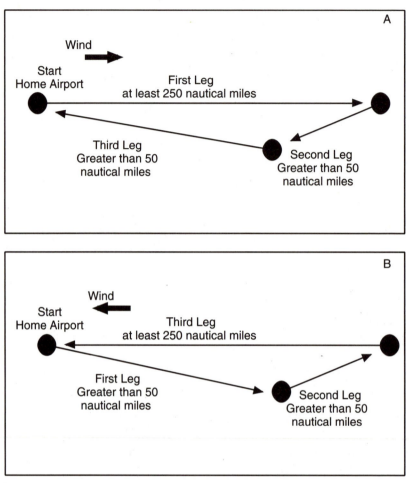

Fig. 11-2 *Diagrams of "long" cross-country requirement for commercial pilot rating.*

These are questions that air taxi and corporate pilots must answer every day, so facing these questions as a prospective commercial pilot is excellent real-world practice.

Commercial maneuvers

Making commercial maneuvers applicable to the real world is very hard to do. After all, you will never hear an air traffic controller say, "4601 Romeo, give me a left chandelle out there." There are five maneuvers that are considered to be the commercial maneuvers: The chandelle, the steep turn, the lazy eight, the eight on pylon (pylon eight), and the emergency descent. The steep turn appears in both the private pilot and commercial pilot PTSs, but they are slightly different maneuvers. Private pilots perform the steep turn with 45° bank angle with a ±5° tolerance, whereas commercial pilots must demonstrate a 50° bank angle with a ±5° tolerance. The emergency descent also appears in both the private pilot and commercial pilot PTSs, but the commercial requirement is more stringent, specifically calling the applicant to maintain positive load factors and engine power setting during the procedure. The chandelle, lazy eight, and eight on pylon (pylon eight) are seen only in the commercial pilot PTS.

Some of these five maneuvers are purely pilot show-off maneuvers and have little or nothing to do with real-world flight. Figure 11-3 is a tongue-in-cheek depiction of a realness continuum. Chandelles, lazy eights, and pylon eights belong on the side most unlike the real world. A little more credit is given to steep turns. The emergency descent is the only one that being really good at might actually help you out someday.

I call them show-off maneuvers because they are used to show off a pilot's skills of airplane flying. These skills, if present, will be showcased by the performance of the maneuver. These skills, if not present, will be exposed as lacking by these maneuvers. This is why they are included on the flight test. They are easy to grade. It would be more valuable to test the commercial pilot applicant on such qualities as professionalism or decision making or situation awareness, but these abstract ideas are hard to grade. So the FAA must take the low road. A perfectly flown chandelle does not guarantee that the pilot is a professional. A pilot who can fly an excellent

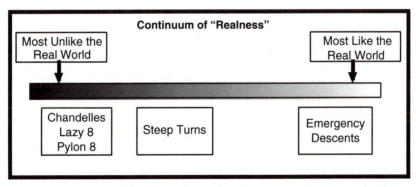

Fig. 11-3 *Diagram depicting how closely certain commercial pilot requirements resemble real-world flying conditions.*

pylon eight is not necessarily a competent decision maker. If I am a paying passenger on an airplane being flown by a commercial pilot, I would rather the pilot be able to maintain situation awareness instead of demonstrating a perfect lazy eight. I was told by an FAA-designated pilot examiner once that commercial applicants who could just spell lazy eight would pass that part—what he wanted to see was if the pilot could be trusted with passengers. That examiner has got a tough job because taking care of passengers is never specifically stated as a requirement for becoming a commercial pilot.

The maneuvers have their place. I am not recommending that they be eliminated from commercial pilot standards, but at least call them what they are: a poor substitute for grading an applicant's ability to be a commercial pilot in command.

Complex airplane

Part of becoming a commercial pilot is the ability to safely fly airplanes that have more sophisticated systems. The commercial pilot PTS further defines what type of airplane must be used:

> *The commercial pilot applicant is required by 14 CFR Part 61.45 to provide an airworthy, certificated aircraft for use during the practical test. This section [of the practical test standard] further requires that the aircraft:*
>
> **1.** *Have fully functioning dual controls...;*
>
> **2.** *Be capable of performing all appropriate tasks for the commercial pilot certificate or rating and have no operating limitations that prohibit the performance of those tasks; and*
>
> **3.** *Must be a complex aircraft furnished by the applicant for the performance of takeoffs, landings, and appropriate emergency procedures. A complex land plane is one having retractable gear, flaps, and a controllable propeller or turbojet propulsion.*

On August 4, 1997, the law changed. High performance and complex are now different. FAR 61.31(e) defines complex and 61.31(f) defines high performance. When the law was changed, the FAA explained the need for the change this way:

The FAA believes that the operating characteristics of complex aircraft and high-performance aircraft are so different as to justify separate endorsements. There are now turbine-powered aircraft that are high-performance but that are not considered complex aircraft. Also, training in one type of aircraft does not necessarily transfer to training in another type of aircraft.

Specifically FAR 61.31(e) states:

No person may act as pilot in command of a complex airplane (an airplane that has a retractable landing gear, flaps, and a controllable pitch propeller) unless that person has

(i) *Received and logged ground and flight training from an authorized instructor in a complex airplane and has been found proficient in the operation and systems on the airplane; and*

(ii) *Received a one-time endorsement in the pilot's logbook from an authorized instructor who certifies the person is proficient to operate a complex airplane.*

And FAR 61.31(f) says

No person may act as pilot in command of a high-performance airplane (an airplane with an engine of more than 200 horsepower), unless that person has

(i) *Received and logged ground and flight training from an authorized instructor in a high-performance airplane and has been found proficient in the operation and systems on the airplane; and*

(ii) *Received a one-time endorsement in the pilot's logbook from an authorized instructor who certifies the person is proficient to operate a high-performance airplane.*

Some airplanes can be both complex and high performance. Others are complex only and still others are high performance only. But for purposes of the commercial checkride the applicant must show up with an airplane that meets the complex definition.

To qualify to take the commercial checkride in the first place FAR 61.129(a)(3)(ii) Aeronautical Experience—Commercial Pilots requires

"Ten hours of training in an airplane that has a retractable landing gear, flaps, and a controllable pitch propeller, or is turbine powered" and this training must be stipulated with an instructor endorsement.

With more systems comes a greater need to troubleshoot. There is much more to do during preflight inspection and runups. The checklists start to become really vital. Although all pilots should use checklists, there are fewer items on fixed-gear, fixed-pitch airplanes than on complex airplanes. There are many more emergency scenarios when you have more systems, and each of these will have an independent checklist. Understanding how the complex airplane's systems work and becoming very familiar with the checklists will ultimately provide a foundation for decision making in normal operations and in a crisis.

Commercial pilot operations

The real world of commercial pilot work is very different from the training environment that pilots start in. When in the training environment, the pilot is nurtured, treated like a valued customer (most places), and watched over. The pilot's first job, being paid to fly, is a real eye opener. No longer are pilots protected and valued; they are employees. The boss will not care much about their problems. In the boss's view, pilots are there to do a job, and if they cannot get that job done, the company will get someone else who can. In the training environment safety is taught and practiced as if it was the only thing. In the commercial pilot work world, safety is very important, but the dollars, the profit and loss, and the bottom line are always stressed and hang over a pilot's head. Every day pilots are faced with safety margin versus money trade-offs. Green and direct from the training environment, this world takes some getting use to.

Are the pilots who are being trained to earn the commercial pilot certificate aware of what lies ahead? I was not. In fact, I can remember being told that the commercial pilot checkride was just a "big private," meaning that the commercial was just a redo of the private test but with a few extra maneuvers thrown in and completed in a bigger airplane. It is important for instructors not to teach the commercial like it was a big private and to spend several hours discussing the huge responsibility difference between the private and commercial certificates.

To illustrate the difference in responsibility between the two, examine these two situations.

Scenario 1: A private pilot is at the counter of his local fixed base operator (FBO) and is in the process of checking out a rental airplane for an hour's flight. A friend of the pilot, who was just hanging around the airport, notices that her friend appears to be preparing to fly and asks, "Hey, can I ride along with you?" The pilot says, "Sure come along, I'm practicing chandelles and lazy eights today." The two go off to the airplane and fly.

Scenario 2: A newly hired commercial pilot is hanging around the FBO when the telephone rings. The pilot answers the phone and talks to a woman who wants to hire a pilot. The woman wants to fly over the city and take photos of the house she is building. The commercial pilot says, "I can help you with that, when can you come out?" The two agree to meet later that day. The woman shows up at the appointed time, camera in hand, the two meet for the first time, and they go flying.

Now let's analyze the differences within these two scenarios. In the first story the person who asked to ride along was an acquaintance of the pilot. Being acquainted with the pilot, she had the opportunity to form an opinion of the pilot beforehand. If she thought that the pilot was not a safe pilot, she would not have asked to ride along. But because she had the benefit of knowing the pilot she could make the judgment call that it was safe to fly with that pilot. The second story was different. The two people in the story were strangers. The woman with the camera did not know the pilot and therefore could not make a judgment call as to the safety of the pilot. The photographer went flying and placed her entire safety in the hands of someone whom she had just met. What assurance would the photographer have that she could trust this pilot with her life? The assurance is the commercial pilot certificate. The commercial certificate is the accreditation of the pilot that says "This pilot has gained the experience and is ready to take on the additional responsibility of ensuring the safety of the general public when they fly."

I moved to a new city once and had to change churches, barbers, and eye doctors, among other things. When I went to the new eye doctor for the first time, I was a little nervous. I was going to let a complete stranger poke around my eye. How could I be sure that this doctor was qualified and proficient? I did not know the doctor, so I had to trust the licensing process. The commercial pilot certificate is the same thing. The unsuspecting general public must trust that the licensing agency (in our case the FAA) has done its job and has not let any unqualified or unproficient pilots through the screen.

The commercial pilot certificate carries with it this additional trust. It is not a big private because there is a greater burden of responsibility. Unfortunately, once again, the commercial pilot practical test little emphasizes this responsibility and instead uses more maneuvers in an attempt to judge the pilot worthy of being a commercial pilot. This is a glaring example of the gap between maneuvers- and mission-based flight training.

Emergency readiness

When I fly as a passenger, I know that the commercial pilot flying the airplane is overpaid on most days. But on the day when all hell breaks loose he or she will be underpaid and in the end it all averages out. From the law's point of view it is one thing to go out and risk your own life, but it is quite another to risk the lives of others. This makes it incumbent upon the commercial pilot to be proficient, remain aware, and deliver safe operations even under the threat of emergencies.

The commercial pilot PTS outlines 14 specific emergency occurrences when the pilot must be able to "analyze the situation [that is creating the emergency] and take appropriate action." The 14 items are

1 Partial power loss
2 Engine failure during various phases of flight
3 Engine roughness or overheat
4 Loss of oil pressure
5 Fuel starvation
6 Smoke and fire
7 Icing
8 Pitot/static/vacuum system and associated flight instruments
9 Electrical systems
10 Landing gear
11 Flaps (asymmetrical position)
12 Inadvertent door opening
13 Emergency exits opening
14 Any other emergency unique to the airplane flown

The PTS says that from this list the examiner is to test the applicant on at least five of the items using simulations.

Scenario: A pilot enters the standard traffic pattern at an uncontrolled airport. Following the proper procedure, the pilot runs through the prelanding checklist. As a part of this routine check the pilot verifies that the airplane is flying slower than the published landing gear operation speed and selects the "gear down" position of the gear handle. Nothing happens. The landing gear does not come down. The landing gear position indicator does not show a gear-down indication. What should the pilot do now?

All pilots who fly a retractable landing gear airplane should play this scenario out over and over in their minds. First, leave the traffic pattern, climb, and get some space where you can think. Even though you should know it already, get out the landing gear emergency checklist and go through it step by step. The emergency gear extension procedure will almost always succeed in lowering the landing gear. When the gear comes down and/or a down indication is received, make a normal landing. What if the emergency procedure fails to bring the landing gear down? Knowing how the landing gear system works will be essential now, but remember, you are never alone as long as you have a radio. Even at an uncontrolled field there can be help. Call unicom and tell whoever answers to put an A&P technician on the radio. The technician may have some ideas or remedies that you might not have thought of. Flying high G turns has been tried. Adding fluid to a hydraulic reservoir might work. It will depend on the system. All during this time keep your passengers advised, but don't say anything that could make them more nervous than the situation already warrants. Don't say, "This gear problem really isn't a big deal, very few of these turn into fatal accidents!"

If all efforts to lower the landing gear fail and a gear-up landing becomes inevitable, there will be an even greater need to seek help. Fly past the runway with people watching so that you can know if all the landing gear is up or if some are up and some are down. This will play a big role in deciding your course of action. You might want to fly away from the uncontrolled airport to another airport that has long runways and rescue help. Your first priority is to protect the safety of yourself and your passengers. The airplane comes second. This means that you should not place additional hardship on the approach and landing by taking exceptional means to shut down the engine or engines and bumping the starter in an attempt to place the propeller in a horizontal position. I am not saying that you should never attempt this, but you would want a very long runway. You do not want to overshoot or undershoot the runway because you were preoccupied with saving the equipment. A pilot was held in violation and had his certificate suspended once for what the FAA determined was placing equipment safety above passenger safety.

The incident involved a cabin-class twin-engine airplane. There were six passengers and the one pilot aboard. During the preparations for landing, only the main landing gear came down. The pilot could not get the nose wheel to lower, so he had all the passengers leave their seats and move to the rear of the airplane. His plan was to land nose high. With the weight concentrated in the rear of the plane, he would attempt to hold the nose off the runway as long as possible. He hoped that when the nose dropped and the propellers began to hit the ground, the speed would be reduced to such a degree that minimum damage would result. The plan worked. The airplane suffered minimal damage, but the passengers were sitting all over each other in the aft baggage hold where there are no seat belts. The FAA believed that it would have been better to leave the passengers belted in their seats for their safety than to shift the weight to the rear for the airplane's safety.

When you begin to fly a specific type of retractable landing gear airplane, have this gear-up scenario discussion first with a flight instructor and then with the A&P technician who works on the airplane. They will give you several ideas to try on that particular airplane if and when this ever actually happens.

Scenario: A young commercial pilot flying during the first week as a new employee at a small corporation, lands and lets off the company's CEO and two vice presidents. The three company executives leave in a cab for a very important meeting, leaving the young pilot at the airport for the majority of the day. The pilot makes herself comfortable in the crew room, but soon daytime TV and old magazines get old, so she steps outside to the airplane. She orders fuel for the return trip that afternoon and then notices something peculiar. The needle of the outside air temperature gauge has broken off from its stem and is lying inside the instrument's glass cover. What should the pilot do about it? Is the airplane safe to fly? Is the airplane legal to fly? Does the pilot have to purchase another OAT from the FBO? What if they don't have the exact same type of OAT in stock? If they don't have another OAT, should the pilot call the home base and have one flown over before the executives return? Will the executives have to stay in a hotel tonight even though they hired the pilot to get them home? Is this the pilot's last day working for this company?

Maybe. If the young pilot cancels an executive's flight when she should not have, she will be looking for work. If she accepts a flight when she should not have, she could be looking to reinstate her pilot certificate. The fact that the pilot is new to the company may be key to the scenario. What kind of indoctrination did this company conduct with this newly hired pilot? At some small corporations there is hardly any flight department and possibly zero formal new hire training.

Here is how to handle the OAT or any other inoperative equipment. First determine if the airplane has a minimum equipment list (MEL). This should have been discussed when the pilot was first hired, but if not, if the airplane does have an MEL. It must be inside the airplane. The MEL must be accompanied by a letter of authorization from the FAA that lets that particular airplane operate by its own rules: the MEL rules. If the airplane has an MEL, the pilot should look through its pages until an entry for the OAT is found. There will be instructions there about how exactly to proceed. The instructions are called M&O procedures. If the fix is an M procedure, it will require a maintenance technician to do the job. If the fix is an O procedure, operations (the pilot) can do the job. The rest is simple; just do what the instructions say and make any record entries. It would not be a bad idea to call back to the home base and inform the chief pilot and/or chief of maintenance what you are doing.

What if there is no MEL? In that case you fall back to FAR 91.213(d)(2), which is the rule for flight in an airplane that has some equipment inoperative. The regulation essentially spells out a four-step test that can be applied to determine if the airplane is legal to fly. The four-step test says that an airplane can be legally flown if

> *(2) The inoperative instruments and equipment are not*
>
> > **(i)** *Part of the VFR-day type certification instruments and equipment prescribed in the applicable airworthiness regulations under which the aircraft was type certificated;*
> >
> > **(ii)** *Indicated as required on the aircraft's equipment list, or on the Kinds of Operations Equipment List for the kind of flight operations being conducted;*
> >
> > **(iii)** *Required by 91.205 (the instrument and equipment requirements for United States aircraft) or any other rule of this part for the specific kind of flight operation being conducted;*
> >
> > **(iv)** *Required to be operational by an airworthiness directive.*

To get your hands on an actual VFR-day type certificate, speak to your A&P technician. The aircraft's equipment list and the kinds of operations equipment list mentioned in (ii) should not be confused with an MEL. The lists referred to in (ii) are located in the airplane's *Pilot Operating Handbook* (POH) or *Airplane Flight Manual* (AFM). The mention of "any other rule of this part" is the catch-all phrase. For instance, if the failed instrument is a turn coordinator, I can fly VFR, but not IFR because a rate-of-turn indicator is required for IFR under FAR 91.205(d)(3). I could fly without a Mode C transponder as long as I was not going anywhere that required one (above 10,000 feet, Class C and B airspace). I could fly without oxygen on board, but not if I planned on flying above 12,500 feet for more than 30 minutes. You get the idea. Part (iii) of this rule becomes a real regulations refresher course. The last item on airworthiness directives (AD) is also worth a conversation with the A&P who services the airplane. An AD is like a recall. If a part has been known to fail on a particular airplane, a letter is sent out to the owners of that model airplane. The AD might be a one-time fix or a recurring inspection that will come due periodically over the life of the airplane. These ADs might require that equipment to be operating for flight.

If the inoperative item in question passes all four tests, you can fly, but not until you follow the rules of FAR 91.213 (d)(3) which says

> *The inoperative instruments and equipment are*
>
> **(i)** *Removed from the aircraft, the cockpit control placarded, and the maintenance recorded in accordance with FAR 43.9 of this chapter; or*
>
> **(ii)** *Deactivated and placarded "Inoperative." If deactivation of the inoperative instrument or equipment involves maintenance, it must be accomplished in accordance with part 43 of this chapter.*

So you cannot fly even if the broken equipment passes all four tests until you take the equipment out or you put a sign on it that says "Inoperative." Now, if you take the item out of the airplane, it may mean that an alteration is needed on the airplane's weight and balance form. If any weight for the item is listed on the form, a weight and balance revision will be necessary. Any change of the weight and balance form requires an A&P technician to sign off on the change. The FAA in further explanation has said that the inoperative placard can be accomplished by writing the word in letters that are at least ¹/₈-inch high, on a piece of paper or tape and attached to the inoperative item.

As you can see, it is good that the pilot has the rest of the day to figure all this out. But are we sure that it is the pilot who can conduct the four-step test? Contrary to arguments that I have had with FAA maintenance inspectors, the regulation is clear that the *pilot in command makes the call.* FAR 91.213(d)(4) says, "A determination is made by a pilot, who is certified and appropriately rated under part 61 of this chapter...."

This gets the pilot home with the broken OAT gauge. The CEO and the two vice presidents never know the difference, and the pilot looks like a pro to the chief pilot and the chief of maintenance the next day. Once the airplane gets back into the hands of maintenance personnel, the burden of decision shifts to the A&Ps. FAR 91.405 says that any inoperative item that passes the four-step test and is flying around in an airplane must be "repaired, replaced, removed or inspected at the next required inspection." When the next inspection comes due, it will be the maintenance technician's responsibility, not the pilot's, to return the airplane to service. The technician may elect to leave the item broken but must make a determination first that the problem still poses no threat to safety and is not adversely affecting any other part or aircraft system. If items are not repaired or replaced, the technician must ensure that the placards are still correct and that entries have been made in the maintenance records returning the aircraft back to service.

There is no crossover. If you have an MEL, it must be used. If in the field a pilot utilizes FAR 91.213 instead of the MEL to get home, the MEL must be surrendered back to the FAA. Regulation 91.213 applies to all pilots of any grade. Do not let this commercial pilot scenario fool you into believing that this does not apply to all pilots.

I think that telling a story to pilots about another pilot in a jam is more effective than having the pilot memorize a bunch of regulations. Once again, people make better decisions when they have been trained to make decisions. Memorizing regulations (even if you could find the regulations that apply to a circumstance) is like memorizing a maneuver. Providing a scenario as the basis for discussion changes the process from a maneuver- to a mission-based (How am I going to get home?) learning experience.

12

Decision Training
for Multiengine Pilots

Becoming a multiengine-rated pilot does not take very long, but it is filled with peril. The training is usually one emergency after another (Fig. 12-1). Multiengine airplanes obviously cost much more money, so there is always a sense of urgency when training in them. Most multiengine courses are accelerated and geared for efficiency. Because of the additional cost, multiengine instructors try to pack a great deal into each multiengine flight. This is great for cost savings, but it also means that the lesson jumps from one scenario to the next with little time to understand the real-world implications. For that reason, it pays (in dollars and time) to ask dozens of what-if questions outside of the actual airplane. Multiengine training can otherwise boil down to a series of emergencies and rote remedies for those emergencies. This makes multiengine training very maneuver-based, and it seems that the rapid pace of the training leaves no time for scenario-based work. As with the three previous chapters, this collection of multiengine training scenarios is not a complete set. There are many excellent examples beyond what is presented here that creative instructors and pilots can use once they start using real-world situations. Also, no attempt is made here to explain multiengine topics; rather it applies the topics to actual situations.

Multiengine aerodynamics

Scenario: A pilot takes off in a multiengine airplane. Just as the airplane reaches 300 feet AGL and the runway passes behind, the left nose baggage door pops open. A suitcase comes flying out of the baggage compartment and strikes the left engine. This damages the left propeller and thrust is lost. What does the pilot do now?

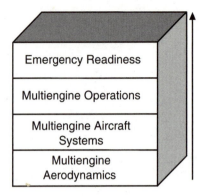

Fig. 12-1 *Multiengine pilot building blocks.*

Should the pilot attempt to climb away using the right engine, crash land into whatever lies ahead, raise the landing gear, feather the propeller, apply rudder or do all of these?

The airplane is low. Climbing to safety on the remaining engine would be wonderful, but believe it or not, that is not the first concern. The first concern must be with keeping the airplane under control. If the right engine causes the airplane to yaw faster or with more force than the airplane's right rudder can counteract, the pilot will lose complete control of the airplane. The airplane will then yaw-roll and soon thereafter strike the ground, probably inverted. This is that last thing any pilot wants. Many pilots believe that just because they have an "extra" engine, they can climb to safety, but this is a myth. A crash landing while under control is better than an attempted climb where control is lost.

Part of learning to fly multiengine airplanes is learning about the aerodynamics that create airplane control. When a decision is warranted in multiengine flight, there will be absolutely no time to deliberate. The decisions must be made almost instantaneously with the problem that created the decision circumstance in the first place. If there was ever a situation that demanded asking the what-if questions, this is it. What if one engine quits during the takeoff roll? What if an engine quits after lift off with runway ahead? What if an engine quits with no runway ahead? All these questions must be worked out in advance. Consult the control speeds and performance numbers of the multiengine airplane you fly to get the particulars, but ask real-world questions before your next takeoff.

Multiengine aircraft systems

Scenario: A pilot lifts off the runway flying a light twin-engine airplane. The left engine fails just as the end of the runway passes underneath the airplane. Higher terrain is just ahead. When the left engine quits, the right engine, which is at full power, yaws the airplane to the left. The left engine is dead and the right engine is alive, so the right side of the airplane is trying to outrun the left. The pilot uses right rudder to stop the yaw to the left. The pilot struggles to bring the airplane under directional control. The left engine is no longer providing thrust, but worse, it is now a source of drag. It is taking more and more rudder to keep the airplane under control. The pilot could reduce power on the right engine to reduce the yaw, but the rising terrain ahead requires a climb. What now?

The dead engine is no longer producing power, but that does not mean that the propeller has stopped turning. In fact, looking at the left engine, it appears to be working properly. The airflow through the propeller arc is pushing the propeller blades at an angle that makes them "windmill." In other words, instead of the engine driving the air, the air is driving the engine. As the lifeless propeller turns through the air it creates drag across the entire "plate" of the propeller disk. Figure 12-2a illustrates the windmilling propeller pulling back while the operating engine is pulling forward. These two forces act on the pivot point, or center of gravity, of the airplane, and as you can see, this airplane is about to be flung like a Frisbee. To reduce this tendency, the propeller systems of multiengine airplanes must have a feature that single-engine airplanes do not have. The propeller blade must turn edge-on to the wind, into what is known as the "feather" position. Figure 12-3a shows the propeller blade windmilling and producing high drag. The drag is not only produced across the blade as if it were stationary, but drag is created over the entire arc of the propeller. Drag is produced over the shape of a disc. When the blade is turned edge-on (Fig. 12-3b), the relative wind does not drive the propeller into a windmill because the majority of the relative wind misses the blade altogether. This greatly reduces the drag and lowers the flung-like-a-Frisbee effect (Fig. 12-2b). The pilot will better be able to control the yaw with the rudder that is available. In this scenario, the pilot should quickly feather the left engine's propeller and maintain directional control with right rudder.

When you check out in an multiengine airplane, one of the items that must be fully covered and understood is how the propellers feather. The changing blade angle is not unique to complex airplane pilots, but a single-engine constant speed propeller does not go all the way to the feather position. The mechanics of how it is done are unique from one airplane to the next, so understand how yours works so that you can answer the what-if questions.

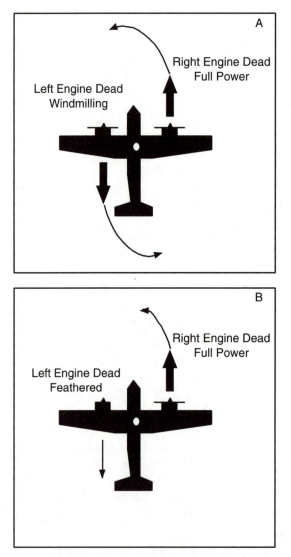

Fig. 12-2 a *Depiction of the effect of a dead engine windmilling.* b *Depiction of the effect of a dead engine feathered to prevent windmilling.*

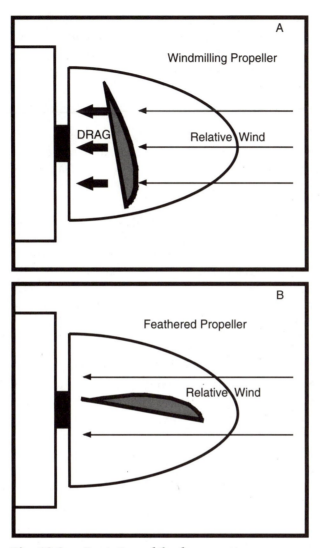

Fig. 12-3 a *Depiction of the forces acting on a windmilling propeller blade.* b *Depiction of the forces acting on a feathered propeller blade.*

Multiengine operations

Scenario: A pilot is flying at cruising altitude in IFR conditions. The pilot has filed to an airport with an NDB approach and has an alternate airport with an ILS approach. The flight has been long, almost 3 hours. Then the right engine begins to lose power. The pilot checks the mixture, the carburetor heat, and the fuel pump, but nothing restores power, and then the engine quits altogether. The pilot follows the proper engine shutdown and securing procedure. The clouds are low, possibly too low for the NDB approach at the original destination. The alternate airport, with the ILS, is 50 miles away into a strong headwind. What does the pilot do now?

The airplane is now in flight on only the left engine. To maintain level flight, a higher than normal power setting is required and fuel flow is higher. Should the pilot fly a single-engine NDB approach at the destination airport? If the clouds are higher than the MDA, a single-engine approach and landing can be accomplished, but if the clouds are lower than MDA, the pilot will be faced with a single-engine climb out and missed-approach procedure. If the pilot shoots the NDB approach, it will be at the expense of fuel, fuel that could have been spent en route to the ILS approach. No matter what, more fuel will be needed for the left engine to fly to the alternate than originally planned because that engine is operating above cruise power setting to maintain level flight. Is there enough fuel in the left wing, supplying the left engine, to reach the alternate?

This situation is a mixture of problems. There are instrument flying questions: the clouds and MDA. There are multiengine aerodynamic questions: single-engine performance and a missed-approach climbout possibility. There are fuel-planning questions: gamble the fuel at the destination attempting an approach or save fuel going straight for the alternate? All these questions are combined into a classic decision scenario. Any decision the pilot makes will alter the options later. This does not present a single simple choice, but an interwoven series of choices. What is the best course of action? I would want more information. What is the cloud ceiling at the airport with the NDB approach? Can I get that information from an AWOS station at the airport? If I attempt the NDB approach, do I leave the airplane configured for a go-around during the approach? Usually the landing gear is lowered at or before the final approach fix, but this time do I leave the gear up until I see the runway? That way, if I do not see the runway, I'm already in a low-drag configuration to help with the climbout. What about the fuel? Whether or not I attempt the NDB approach first, can I make it to the alternate airport? The right engine is dead, but the right wing still has fuel in it. Can I bring the fuel from the right side of the airplane to extend the range of the left engine? Multiengine systems, weather, IFR operations, and single-engine aerodynamics must all be considered, and considered fast.

Emergency readiness

Each of the previous scenarios involved some sort of unusual or emergency situation. Multiengine training is dominated with preparing for the worst. When multiengine training begins, try to fly the airplane just as you would have done with a single engine. Use both throttles as if they were one. Except for slightly more speed, fly the instrument approaches just like a single engine. This "fly-like-a-single" approach will quickly make you feel comfortable in the twin, but this part of the training will not last long. Because the length of training time is short, driven by the higher cost, the demands will quickly shift to multiengine-specific problems. This means emergency training, and that is why all the scenarios of this chapter have involved emergencies.

Scenario: A pilot flying a light twin-engine airplane has experienced an engine out on takeoff. At the time of the engine failure the airplane was flying fast enough to maintain directional control. The pilot quickly and efficiently reconfigured the airplane to produce the best possible single-engine climb performance. The airplane is at or near its maximum takeoff weight, and it is a hot day so it seems to take forever, but the airplane reaches 500 feet AGL. The pilot turns back toward the airport. Once in level flight the pilot reduces power on the good engine and sets up a low traffic pattern. On final approach to the runway, the landing gear is down, and power is reduced on the good engine. At 200 feet AGL and on one-quarter mile final, a herd of deer runs out on the runway. What does the pilot do now?

If the pilot elects to make a single-engine go-around, the power must be brought back up to full power on the good engine. This will reintroduce the severe yaw that threatened this flight to begin with. The pilot could only hope to maintain directional control if speed is held constant or even increased. The landing gear and flaps will need to come back up to reduce drag. And all this must take place during a time when speed and altitude are both reducing.

The other option is to land anyway, despite the deer on the runway. If this option is taken, every effort should be made to miss the deer, even if that means landing on the grass next to the runway. During every single-engine approach of a multiengine airplane, there is a point of no return. Reconfiguring the airplane down low, transitioning from a descent to a climb with only one engine and incurring the inevitable control problems at slow speed, will be at least as risky as landing straight ahead. The altitude that should be considered the single-engine approach point of no return will depend on the airplane's single-engine performance, the loading, and the density altitude. Every multiengine pilot should work this out in advance because there will be no time as the deer are running out of the trees to calculate density altitude. In most light twins of the type in which initial multiengine training is conducted, you are committed to land no matter what when you are below 200 feet on a single engine.

These scenarios require more background knowledge than was presented here. The factors involving Vmc, accelerated-stop distance, and multiengine systems would all play a major role in arriving at a final decision. But as with any time-crunched decision, the more information at hand and the more familiar the pilot is with the circumstance, the better the decision and outcome.

13

Recommendations

At the conclusion of most books all the loose ends are neatly tied up and a sense of closure is realized. Not this book. The observations made were like turning the spotlight back on ourselves in general aviation. Both great exceptions and worst fears were realized by looking in the mirror. I believe that the ideas, concepts, and discoveries of this book, if implemented, would improve and also fundamentally change our general aviation flight training. The following are recommendations to various groups within aviation to help with implementation.

Recommendations to flight instructors

1 Do not "script" the flight lesson. If multiple instrument approaches are needed for student practice, use a variety of outcomes: missed approach and landings. Make the students decide on a missed approach or a landing every time. Do not let the decision be made for them. When the controller asks, "How will this approach terminate?" ask if you can get "the option" (see Chap. 7). When the controller approves the option, the scenario is back in your hands. You can land straight ahead, you can execute the missed approach, you can even circle to land if the runway you have the option for is not a straight-in runway. Play the option card, protect your scenario, and keep the decisions in the hands of the pilot.

2 Plan instrument lessons in segments. Fly the first segment to an airport other than the airport where the takeoff was made. Force the controllers to treat the instrument training flight as they would an IFR en route flight. In the real world, airplanes take off and travel to their destinations. Airplanes never take off intending to come right back to the place they started. But in flight instruction, lessons almost always start and end at the same location.

When a flight leaves an airport and returns to the same airport, controllers will automatically know the purpose of the flight is training and if the weather is not actually IFR, they will not provide real-world service. This presents an unreal situation to the student. When students become accustomed to the unreal flight training environment, they have problems in the real world when they are exposed to it after training. By flying segments, the training flight becomes an itinerant flight. The *Aeronautical Information Manual (AIM)*, paragraph 4-3-21, designates itinerant pilots with a higher priority over training pilots: "The controller will control flights practicing instrument approaches so as to ensure that they do not disrupt the flow of arriving and departing itinerant IFR or VFR aircraft." This statement taken directly from the AIM is a root problem. How can pilots in training ever learn what it is like to be an itinerant pilot in the real world if they are systematically singled out and given nonitinerant treatment when in training? There is not a better example of the gap between the real world and the training world than this. So savvy instructors must become itinerant to receive real-world itinerant handling from the controller. File IFR to a destination other than the airport at which the flight originates. Teach in segments. Force the controllers to be part of your simulation. Be itinerant.

3 Ask for and expect further clearance times for holding patterns, even in VFR conditions. When flying an IFR simulation flight in VFR conditions, the controllers may not offer the EFC as a part of the clearance. Use good judgment when controllers are overloaded with VFR traffic. It might not be possible to get an EFC in VFR conditions, but in this situation make sure that instrument students understand that in "real" IFR circumstances an EFC time is a holding-pattern requirement.

4 Incorporate real-world decision scenarios into flight training. Have students make decisions about all aspects of the flight. Make decision making a common, not a rare, practice. The examples in this book are just that. Instructors are capable of developing other real-world problem solving for their students. Use the examples of this book to get started with your own ideas. Become a storyteller.

5 Utilize flight training devices (simulators) and personal computer aviation training devices (PCATDs) to expose flight students to real-world scenarios when they are not practical in actual flight.

6 Make it a standard practice to visit a radar facility with every student, and especially every instrument student, so that the student can see the radar screen in action. For VFR students, go

when VFR flight operations are in progress. For IFR students try to make this visit when instrument conditions actually exist. This will help the student visualize the position of the airplane when in flight. Also, controller-pilot myths and barriers tend to come down when the actual people meet face to face.

7 Flight instructors should discuss with students the fact that circumstances arise where the flight instructor plays the role of the controller to simulate certain flight situations. Without a discussion of this role playing element of the training, students can believe that role playing is not taking place and get a false impression of the real-world environment.

8 Instrument flight instructors should stay current and proficient on IFR at all times. Then take instrument students into actual IFR conditions whenever practical. This eliminates many training environment problems. The clouds are a better trainer that the hood can ever be.

9 Meet and work with the controllers in your area. Invite a controller to ride along in the back of an airplane while you conduct a lesson involving air traffic control communications or IFR simulation. Make the controllers understand that they are part of flight instruction and therefore that they are partly responsible for developing pilots who can safely function in the ATC system.

10 When an airplane that you instruct in goes down for an annual or 100-hour inspection, go with it. You will be able to teach systems more effectively when you yourself see the systems first hand. Develop a rapport with the maintenance technicians who do the work on your training airplanes. Start a collection of discarded parts that can be used in your own systems demonstrations. Take students over to the maintenance facility so that they, too, can see first hand what makes these systems work. I saw a sign hanging in an aircraft maintenance facility once that said, "Shop rates: $50 per hour. $75 per hour if you watch. $100 per hour if you help." It is hoped that you have a maintenance staff that does not have this attitude. If you do, you need to start cultivating a better working relationship.

11 Do not perpetuate myths about declaring emergencies. Help students understand that an emergency can be a life-threatening situation and no paperwork or questioning is more important than the safety of the flight. Given a choice between declaring an emergency and thereby eliminating a safety threat or remaining quiet and remaining in a dangerous situation, declare the emergency every time.

Recommendations to flight students

1 Students should understand the roles being played during VFR and IFR training. Sometimes the flight instructor will give instructions as if they were the air traffic controller to facilitate the simulation of some flight situations. Students should always be aware that this is only role playing and understand how that would have played out in a real-world situation.

2 Students should realize that the goal of every instrument approach is to make a landing. If you are asked to execute a missed approach even though the runway may be visible, this is only done to get as much practice into a lesson as possible and save you money. It should not replace the need to make a missed-approach versus landing decision.

3 After each instrument flight lesson the students and instructor should recreate what happened during the flight. Discuss what the controller's instructions during the lesson were and note instances where what was said might have been different if the flight had been conducted in actual IFR conditions. Students should ask questions if they are unclear what these differences are.

4 Expect to make decisions. Learn to be assertive. Do not be so afraid of doing the wrong thing that you do nothing at all.

5 Students should learn that they are training to be in command and that controllers work for you. The pilot has the authority to take control of any situation, especially if that situation involves an emergency.

6 Students should understand that declaring an emergency is an option that can be used without fear of FAA repercussions. It is better to call the tower and explain how and why you did something after landing than to place you and your passengers at greater risk in the air by remaining silent.

Recommendations to air traffic controllers

1 Controllers should understand that they are partners with the flight instructor in training decision-capable pilots.

2 Controllers should handle an incoming Boeing 737 airliner and a Cessna 172 on a third consecutive approach with the same phraseology and temperament. The controller should understand that the pilot of the 172 may be in training to become the pilot of the 737. The training will not be complete if the ATC handling is not the same.

3 Controllers should work with area flight instructors to develop training protocols. There may be times and places where training should not take place due to heavy controller workload. If this is the case, instructors should be informed.

4 Controllers should realize that certain requests made to an aircraft involved in flight training could jeopardize or destroy the scenario that the instructor has created. Asking a pilot, "Will this be a low approach or full stop landing?" may be necessary to plan separation, but controllers should understand that asking this question takes the decision away from the pilot in training. This, over time, does not teach the pilot to make the decision when in actual IFR conditions. Methods should be developed so that these question are limited or eliminated. I have never heard a controller ask, "United 254 will this be a low approach or full-stop landing?" The training community does not have a "me-first" attitude. We also understand that training aircraft are by and large slower aircraft, which require sequencing, but ATC handling should not be different just because it is a training aircraft.

5 Controllers should be aware of when pilot workload is at its highest level. Controllers should limit nonessential instructions during high pilot-workload segments. When the pilot is making an intercept to the final approach course, it is not a good time to give a new missed-approach instruction.

6 Controllers can sometimes be too helpful. When intercepting an inbound course, the pilot needs a 30 to 45° angle of intercept. The inaccuracy of the equipment, especially an NDB, requires that the pilot, not the controller, calculate the intercept. Controllers, in an attempt to give excellent vectors, sometimes try to turn the airplane directly onto an inbound course with a very narrow (10° or less) intercept. This will often cause confusion to the pilot at a time when the workload is already high.

7 Controllers should also understand that procedures that today are included in the *Controllers Handbook* and the *Aeronautical Information Manual* can actually insulate the training pilot from the real world. These procedures should be changed and new procedures adopted that encourage pilot decision making.

Recommendations to the Federal Aviation Administration

1 The FAA should review the procedure pertaining to pilots who declare an emergency in flight. Emergencies should not be first recorded as deviations as they are today. A new FAA tracking system is needed for emergency situations. The FAA must ensure that pilots are dealt with using common sense and fairness. In a circumstance where a pilot declares an emergency and later it is determined that he or she has violated some other rule in the process, care should be taken. It is understandable that the FAA will want to prevent that pilot from repeating the offense by using enforcement actions. These actions include pilot certificate suspension, revocation, and/or fines. But by taking enforcement action against that one pilot, many other pilots will be deterred from also declaring an emergency, even if they have no other fault. The end result is the opposite of that intended by the FAA. Taking enforcement action against one pilot in an effort to make flying safer actually produces a situation that is more unsafe for the total population of pilots. This is not to say that pilots should always receive immunity for violations when they are discovered through an emergency investigation, but the FAA should look at the larger picture. The FAA safety program managers, not the inspectors, should conduct the inquiry into situations where emergencies have been declared. "Emergency" inquiry, not "deviation" inquiry should be used when an emergency has been declared.

2 The FAA should begin, using every way possible, to educate pilots that the threat of an aircraft accident is more dangerous than the threat from the FAA. There is a universal attitude among pilots that the result of declaring an emergency is scrutiny of them and their airplane by the FAA. They believe that this scrutiny will not be handled with common sense but rather by bureaucracy. There have been cases in the past where the FAA did not deal appropriately with a pilot in this situation. The stories were told and retold, and the evils of the FAA became more myth than fact. The education process can start at pilot proficiency programs, back to basics videotapes, and in advisory circulars. After the education effort is under way, the FAA in every district must double its efforts to use common sense and fairness with pilots who declare emergencies. The first time another "Bob Hoover" case comes along, all the FAA education efforts will have been wasted. Bob Hoover, the famous stunt pilot, had his medical certificate revoked by the FAA. Even though Hoover was later reinstated, many believed Hoover was mistreated by the FAA and the story has been used as evidence of improper FAA tactics.

3 The FAA should work with flight instructors, flight schools, and college and university aviation programs to develop a decision—

rather than a maneuvers-based certification criteria. Today a commercial pilot is required to perform maneuvers that have no application in decision-making skills. New certification requirements now mention decision training but give no guidance about what is meant by this or how the FAA believes it should be conducted. Today, pass or fail does not depend on decisions made but on maneuvers flown. The training of pilots should shift emphasis to decision training. Maneuvers will always be important and required for pilot certification, but the FAA should move to a general aviation pilot qualification program. Such a program would not be dependent on minimum hours flown but on capabilities acquired. The capabilities would include maneuvers but emphasizes decision making. It is time for general aviation AQP.

14

Becoming the Pilot in Command

When I learned to fly, I heard the phrase pilot in command many times. I grew up in aviation thinking that the decisions that were left up to the pilot in command included such things as deciding when to turn base in the traffic pattern, calculating true course and fuel consumption, and determining which en route altitude to fly. These items are important, but being the pilot in command means so much more.

As you follow your journey of flight, be bold, not tentative. Be assertive, not shy. Do not be afraid to act. I was flying with a student recently and as we returned to our home airport, I introduced this scenario: While flying along the engine rpms drop from the cruise setting of 2400 to only 1600. To make the scenario real, I pulled back the throttle to 1600 rpm and said, "The engine has just swallowed a valve. You can reduce power, but for the remainder of the flight you cannot go above 1600 rpm." This particular student took control. First he trimmed the airplane for minimum altitude loss. He quickly switched to the AWOS frequency. The wind was 140° at 5 knots. This wind would favor runway 18, but we were approaching from the south, and so runway 36 was closer. As we came nearer to the airport, our altitude passed through the traffic pattern altitude. The pilot saw that another airplane was at the far end of the airport in the runup area for runway 18; we did not see or hear any other traffic. The pilot was faced with a long downwind, base, and final to runway 18, but we were now well below the pattern altitude. The pilot had to make a decision. A complete pattern to 18 or a straight-in to 36? The pilot considered the wind, which although would be a quartering tailwind for runway 36, was light at 5 knots. The pilot acted. He called the unicom frequency, "Aircraft holding short of 18, would you hold your position? November 5122B is making a partial power landing on 36."

The pilot of the airplane on the ground responded that she would hold short. The landing was made on runway 36 without any further problems. The pilot had taken control. He had demonstrated what it took to be pilot in command. That scenario was not very complicated, but the student had been placed in a position where a decision was warranted and he made the decision. Like anything else, we get better when we practice. So practice making decisions.

Being assertive means be confident. To gain confidence, read, study, ask questions, and become solid in your knowledge. As your confidence increases, your reluctance and fear of doing the wrong thing will decrease.

Get confidence in your ability to troubleshoot and/or deal with emergencies. Really get to know your airplane. Many maintenance technicians believe that pilots know nothing about the airplane forward of the firewall. Do your part to eliminate that belief. Learn the airplane systems. Talk to your technician. Become an expert troubleshooter who sees the warning signs early and avoids pattern mismatches.

Gain confidence in your pilot and communications skills. Become active in local pilot proficiency programs, PACE programs, and operation raincheck events. And go meet your partners, the controllers.

A pilot in command is an assertive decision maker who lives in the real world. The pilot in command maintains situation awareness. The pilot in command is not afraid to use any resource, and these include controllers, weather briefers, and maintenance technicians. When necessary, pilots in command will take matters into their own hands, declare an emergency, protect their passengers, and get on the ground safely.

One of the pilots who participated in the general aviation pilots study (in the Appendix) may have said it best, "This project may save some lives. Maybe mine."

Appendix

The Pilot Decision-Making Project

I wanted to take a group of general aviation pilots and give them a dose of what airline pilots get. As a part of my teaching position, I work with college students who vie for flight officer internships at three major air carriers. As a faculty advisor I work closely with these air carriers and over time have been in a unique position to see their operations from the inside. I have sat in on many airline training sessions. Some have been for newly hired pilots, and some have been recurrent training for veteran pilots. Some have been in a classroom and some have been in the flight simulator. I have been able to see first hand how these airlines train, and being a general aviation flight instructor as well, I could see the similarities and the stark differences in the two world's teaching methods. So after several years of airline indoctrination, and after seeing the difference in accident trends, I wanted to try something new. I wanted to take the real-world approach that seems to have been proven by the airlines and test it out on general aviation pilots. In the fall of 1997 I did.

I gained access to two flight training devices (simulators) at the Aerospace Department at the university where I teach. I pitched my project to the university's research committee, which weighed all the ethical concerns and later approved the project for use with human subjects. My local FAA safety program manager agreed to give me the names and addresses of pilots living in the three counties surrounding where I live. When I received the list, I had to decipher the FAA code. The names in the database were coded with information as to what level of pilot certification each pilot held. The range of experience inside the population of general aviation pilots was too large to handle as a whole. So I narrowed my focus to "nonexpert" instrument-rated pilots. All airline transport pilots and flight instructors were eliminated from the list. What remained was a list of 1189 instrument-rated pilots living in the three counties. I sent all 1189 of these pilots a letter inviting them to participate in a free project that involved flying in the flight training device (simulator). The initial letter described the project and offered incentives to the pilots for participation. These incentives included the fact that the work accomplished in the flight training devices (FTDs) could be used by the individual pilots to maintain FAA required instrument proficiency. This proficiency is obtained ordinarily by paying a flight instructor. Participation in the project potentially saved individuals up to $200 in flight proficiency costs. Also each participant received a certificate of completion, and attendance at the seminars held within the project were also credited toward the FAA's Pilot Proficiency Awards program, otherwise known as the Wings program, from the FAA safety program manager's office.

Contained in the initial letter was a postage-paid response card. Volunteer participants were asked to return the card prior to a deadline to be included in the project. Response cards that were postmarked on or before the deadline became the pool of participants for the project. After the deadline, the pilots who had responded were contacted by telephone. During the telephone conversation it was verified that the volunteers were aware of the study's requirements and their first session in the FTD was scheduled.

I decided that for the purpose of the project the target would be pilots who fly alone and under IFR conditions. Because single-pilot operations are a fact of life in general aviation, I decided to apply scenario-based, LOFT-style training to the single pilot rather than the flight crew. This decision was made in part because of the gap in the previous work and in part due to the accident statistics. The recognition that this type of flying carries with it a greater hazard and the lack of previous work done with this group made this decision easy. The study therefore had a target population consisting of general aviation pilots who fly IFR as the single pilot.

For my project, FTDs were used because they are closer to the environment in which general aviation pilots train without losing the essential element of realism. Your local FBO is much more likely to

Fig. A-1 *One of the two Frasca 141 Flight Training Devices (simulators) used in the project.*

have an FTD available for use in general aviation flight training than a true flight simulator. I wanted to take a different approach to flight training, but I knew I must remain practical. So I used FTDs.

This project used two Frasca 141 FTDs (Fig. A-1). These FTDs are owned and operated by Middle Tennessee State University in Murfreesboro, Tennessee, and are typical of devices used throughout general aviation. The two FTDs had both been inspected by a principal operations inspector from the Nashville office of the Federal Aviation Administration and found to be in compliance with all applicable regulations.

The project schedule

The plan was to have the pilot volunteers all fly the same LOFT scenario in the FTD. Then after everyone had flown the first session, two seminars would be held. Half the pilots would attend the first seminar and the other half would attend a second one. I told all the pilots that the reason for having two seminars is that I could not get a room big enough to hold all of them at once. That was true, but the real reason was that I gave the groups different seminars. After the seminars, all the pilot volunteers were scheduled back into the FTD for a second LOFT scenario.

On September 2, 1997, 1189 letters to potential volunteers were sent using the FAA database for pilots in three counties of Tennessee. Sixteen of the letters were returned because the pilot no longer lived at the address and had no forwarding address or was marked by the post office as an "undeliverable address." I did not report to the FAA about these 16, but I guess they all had violated the regulation that says you must report any permanent address change to the FAA within 30 days of the change.

One hundred and thirty-nine response cards were eventually returned. The initial letter asked the volunteers to return the response card by September 12, 1997, to be included in the project. Of the 139 response cards returned, 19 were received after the September 12, 1997, due date. Cards were still coming until November 26, 1997. Twenty-five volunteers on the initial telephone contact said that they were unable to attend the seminars that had been scheduled for October 11, 1997, and therefore were eliminated from the pool. The remaining 114 volunteers were randomly divided into two groups, which would become the traditional and naturalistic groups. Time constraints of the project together with several cancellations of participants narrowed the pool from 114 to 72 participants. The loss of 42 participants was very disappointing, but these were real people with real lives that could not be placed on hold just for my project. This let me know that I was not working in a laboratory but in the real world.

Beginning on September 16, 1997, and continuing to October 9, 1997, participants were scheduled into the FTDs and given the opportunity to fly the first LOFT scenario. During that 24-day period 72 participants flew the first LOFT scenario. These 72 participants flew a combined 82.3 hours in the FTDs.

On October 11, 1997, I conducted two workshop/seminars with the participants. From 9 until 11 a.m. the naturalistic group met. From 1 until 3 p.m. on the same day the workshop/seminar for the traditional group was held.

Nine volunteers who participated in the first FTD session did not attend either of the October 11 workshops. These nine were eliminated from the pool. The remaining number of volunteers was then 63.

Following the workshops, the second FTD sessions began. Starting on October 13, 1997, and continuing until November 25, 1997, the participants were scheduled back into the FTDs for the second LOFT scenario. Five participants did not reschedule for the second scenario, so the total number of participants who completed all phases of the project was 58. Ultimately, the naturalistic group had 30 volunteers complete the project and the traditional group had 28. The combined FTD flight time for the second scenario was 75.3 hours. The FTD time for the entire project was 157.6 hours.

The participant's first LOFT session

I met the pilot volunteers one by one and took them to a classroom. I asked them to read and then sign a permission and agreement statement (Fig. A-2). The pilots were asked not to discuss the project with anyone while it was ongoing, and permission was asked of the pilots to use the results of their work in the project for future analysis and publication.

After this paperwork was completed, the pilot participants received a briefing before entering the FTD about the device. I wanted to make sure that every pilot volunteer received the same instructions. Later on I did not want any confusion to cloud the results. I did not want to depend on what I said in the way of instructions. I might accidentally leave something out or inadvertently stress one thing over another. To prevent this, I had each pilot volunteer read the same instructions (Fig. A-3). They were told that they would be placed in a miniscenario involving instrument flight. Once in the FTD and the scenario started, the participant would not receive any instruction and all communication between the participant and the researcher was through air traffic control headsets. I acted only in the capacity of an air traffic controller on an actual flight. All participants were given a brief opportunity to "fly" the FTD in order to become familiar with the device. The first scenario began with the FTD on the ground at the Fayetteville Airport (FAY) in Fayetteville, Tennessee (Fig. A-4 position A). The flight's intended destination was Smyrna, Tennessee (MQY; Fig. A-4 position B). The route of flight was from Fayetteville to the Shelbyville VOR station and then to Smyrna using air traffic instructions called radar vectors. When the pilots finished their "warm-up," they flew past the Shelbyville VOR and at that point I started a timer. Then, following the exact script I began to add elements to the flight that the pilot could not have anticipated. I first

IFR Research Project Permission
and Agreements

Thank you for volunteering to participate in this research project. All information gathered as part of this project will be completely confidential and will be used for research purposes only. This project has been approved by the Middle Tennessee State University ethics committee for human subjects research. The following are the rules for the project.

1. As a participant in this project you need to give written permission for the researchers to use information gathered during the project as a part of a study on IFR flight safety. Although information will be gathered from you, your name will never be used.

2. When in the flight simulator you will be placed in actual IFR flight situations. It is important that information about these situations be kept from other participants in the project so that they can experience the situations first hand and without prior knowledge. We ask that you not discuss this project with other pilots until after the completion of the entire project.

3. As a part of this project, participants will be offered an Instrument Proficiency Check (formally know and an Instrument Competency Check). Our first goal is air safety. With safety in mind there may be some cases, where the time spent in the simulators as a part of this project might not, in the opinion of either the instructor or participant, be adequate to qualify for an Instrument Proficiency Check. In this event the project instructor will recommend additional training beyond the scope of this project. The opinion of the project instructor will be final. For those who do qualify a Instrument Proficiency logbook endorsement will be given at the completion of the second simulator run.

4. All participants must attend the workshop that they have been assigned to on October 11, 1997. If for any reason a participant cannot or does not attend the October 11 workshop they, unfortunately, can not continue in the project.

I understand all the information presented above on this page. I give my permission to use information gathered during my simulator sessions in this research project. I understand and will abide by all rules of this project.

_____   _____
Name   Date

Fig. A-2 *Each volunteer read and signed the IFR Research project permission & agreements document.*

used the computer terminal outside the FTD to cause the airplane's alternator to fail. The failure of the alternator is signified in the cockpit by the illumination of a red light that is marked Alternator.

Soon after the alternator failure, the participant neared the destination airport. The participant was given the Smyrna weather report.

Instructions before entering the simulator.

1. You will be given a few minutes at the beginning of the session to fly the simulator. Keep in mind that the simulator will not "fly" exactly like an airplane. The simulator is for "procedures" training not necessarily "flight skill" training.

2. When the instructor says "lets go" the airplane will be positioned in flight over the Shelbyville VOR at 3,000 feet. You will be flying in the clouds and on an IFR flight plan with the destination of Smyrna.

3. You will need to wear a headset that is provided (or you can use your own). All conversation that takes place between you and the instructor after he says "lets go" must be as conversations between a pilot and an air traffic controller. The instructor is not allowed to give "instruction" but only functions as your controller. You should feel free to make requests of the instructor just as you would an air traffic controller. In actual flight, when frequency changes are made a new voice would be heard. In this simulation, all frequency changes (Memphis Center, Nashville Approach, Smyrna Tower, Nashville ATIS, Nashville Tower) will take place as normal but it will be the same voice. You are to ignore this fact and assume that it is a new controller after each frequency change.

4. You will be given a current set of approach charts (you may choose either NOS or Jeppessen or use your own). Once you begin, you may assume that all frequencies listed in the approach charts are operating properly unless otherwise noted or indicated by OFF flags. This will include ATIS, AWOS, Control Towers, Approach Controls, Center Controllers, VORs, NDBs, Marker Beacons, Localizers & Glide Slopes, etc.

5. Anytime that you fly the simulator to a position that is both below the clouds and in sight of a runway a light will come on in the cockpit. The instructor will demonstrate this light before you begin.

6. Please treat this simulation, in every way, as if you were actually in flight.

7. Good Luck! Have some fun, and thank you for participating.

8. Before you leave confirm the workshop time on October 11.

Fig. A-3 *Each volunteer was given the same written instructions to avoid mistakes or omissions. These instructions were for the first simulator session of the project.*

I was speaking to the pilots over the headset at all times as an air traffic controller. The weather at Smyrna was reported to the participant as "Estimated ceiling 300 overcast, visibility $1^1/2$ miles, wind 360° at 10 knots." The participant was told by me, the controller, to "Expect the ILS runway 32 approach to Smyrna" (Fig. A-5). After the participant intercepted and began to fly this approach, I caused the airplane's glide slope receiver to fail from the outside computer station. At this point there were three elements working together to place the participant in a decision situation. First, the failed alternator meant that all electrical equipment was now being operated from battery power only, with the threat that the battery would run

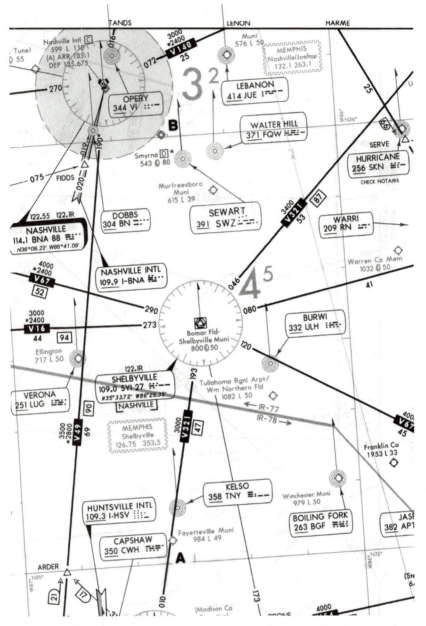

Fig. A-4 *The first session scenario began at Fayetteville Municipal Airport (marked with the letter A in the lower center of the chart) to Smyrna (marked with the letter B in the upper center of the chart).*

down and lose its charge. Without electrical energy to operate the radios, the participant would not be able to fly any instrument approach. In other words, if the battery charge ran out, the participant would be stranded in the clouds with no safe way back down. In this circumstance an emergency is imminent with the possible of a crash likely. So the alternator failure turned the scenario into a race against time. Second, the cloud ceiling was reported to the participant to be 300 feet overcast. This means that the space between the ground and the underside of the solid layer of clouds is 300 feet high. The ILS approach at Smyrna allows pilots to fly to within 200 feet of the ground. This means that a pilot could logically assume that a landing at Smyrna is possible, but the third element changed that assumption. The third element was the failure of the ILS approach's glide slope portion. Without an electronic glide slope capability, the approach at Smyrna only allows the airplane to descend to within 420 feet of the ground. The dilemma facing the participant now is the fact that the approach decent to 400 feet with the clouds down to 300 feet would not allow the airplane to exit the bottom of the clouds. Exiting the clouds is required if the participant expects to see the runway at Smyrna, and seeing the runway is a requirement for landing at Smyrna. I had placed a light in the FTD's cockpit (Fig. A-6) and participants were told before the scenario began that the light would come on anytime they were underneath the clouds and a runway was in sight. They were to assume that they were still in the clouds and unable to see a runway if the light did not illuminate. I had rigged the light so that I could turn the light off or on from outside the FTD. When participants flew the Smyrna approach to the 400-foot level, I did not turn on the light. There comes a point in every instrument approach where the runway must be in sight and a landing accomplished, or a missed approach must be executed. During the first LOFT scenario, if the participant flew the Smyrna approach to the 400-foot level, they did not see the runway and arrived at the missed-approach point without seeing the runway because they were still flying in the clouds. At this point I observed the decisions or lack of decisions that the participant went through to safely land the airplane. I recorded time intervals throughout the scenario, as well as the chronology of the flight past the decision point.

In preparation for the scenario I had spent some time inside the Nashville radar room and control tower. The Nashville radar control tower has the coverage over the area of the flight scenario. I became

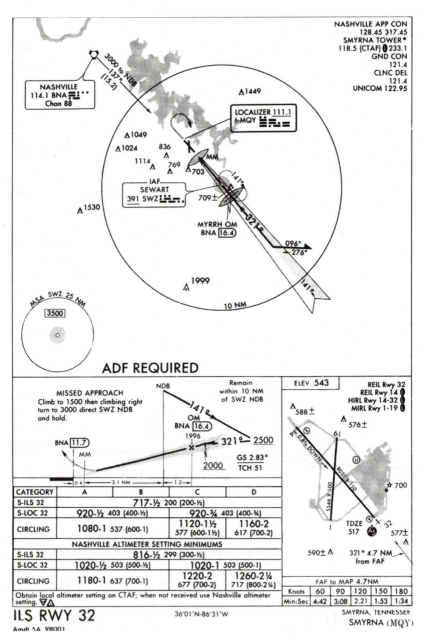

Fig. A-5 *ILS Runway 32 approach at Smyrna, Tennessee.*

Fig. A-6 *Inside the Frasca 141 cockpit, the "night light" is visible. The volunteers were told that a runway was in sight when the light was on.*

familiar with the controllers' procedures for the area, the radar sectors and their boundaries, and the communications frequencies for each sector. I wanted to provide the same service to the pilot participants that they would have received had they flown in this area in an actual airplane.

I also consulted several airframe and powerplant technicians when designing the scenario. I asked them, "How long would a fully charged airplane battery last if the alternator were to quit working?" We calculated the time interval of battery death with all the lights and electronics on and again with a reduced electrical load. During the first LOFT scenario the pilots were given time to work out their problems and end the flight safely based on these calculations. If the pilot left all the lights and electronics on after the alternator failure, they were given less time than if they turned off all unnecessary electrical equipment.

On the first night of the first LOFT sessions, the first pilot flew the scenario beautifully. He understood the implications of the alternator warning light right away. He first turned off and then turned back on the alternator side of the master switch in an attempt to bring the alternator back on line. This attempt failed. I left the alternator light shining brightly. He then turned off the anticollision lights, position lights, and all but one radio. When he received the weather report from Smyrna and later saw the glide slope was inoperative, he understood the math. He knew that if the clouds were actually at 300 feet and he could only descend to 400 feet that he would not see the runway. But he flew the approach anyway, concluding that the 300 feet was only an "estimated" ceiling, and he might fly into a higher cloud base. At the moment he reached the missed-approach point he, without hesitation, executed the missed-approach procedure. When asked by me, the controller, "What are your intentions?" he promptly asked to divert to Nashville (11 air miles away) for an ILS approach there. I gave him vectors to ILS runway 2L at Nashville. When he tuned in the new ILS frequency, the glide slope worked perfectly. In the scenario the glide slope transmitted from on the ground at Smyrna had been inoperative. The weather at Nashville was also reported at 300 overcast, but with a glide slope a descent to 200 feet is possible. This pilot flew the FTD flawlessly, breaking out (I turned on the light) at 300 feet and landing at Nashville with several minutes of battery power remaining. He did very well, but I was crushed. I thought I'd made the scenario too easy. All this planning, all the letters, all the phone calls, and as it turned out this was simply not a challenge. I was not going to learn anything. Then the second pilot showed up.

It was a different story when the second pilot flew the first scenario. He flew the FTD well. He had no problem flying a heading and altitude. But he was unable to fly and analyze problems at the same time. That second session was a wild, sweat-drenching, precrash disaster. It was agonizing to watch, never mind to fly. This poor, FAA-qualified instrument pilot did no planning or troubleshooting and consistently failed to make decisions when they were warranted. He asked the controllers (me) for help and suggestions because he had no idea what to do himself. The whole thing ended when he hit ground level at a high rate of speed with nose low and in the dark because the battery had long since run out of power. I drove home that night thinking, "Well, I have seen the exception," and I was right. I did see the exception that night. But it was the first pilot who had done so well that became the exception, and unfortunately I learned that the second pilot with the deadly performance was closer to the average.

After all the pilot participants had flown the first scenario, 40 percent had experienced a complete electrical failure in flight. In other words, 4 out of every 10 pilots were unable to think their way out of the situation they were in before time ran out. Sixty percent of the pilots did not troubleshoot the electrical problem in any way. They did not recycle the alternator switch, they did not reduce the electrical load, they did nothing. Many later said that they knew the alternator light was a problem, but they did not have enough time together with the other flying duties to do anything about it. Others indicated that they were unaware of what the light being on indicated. They were fully instrument-rated pilots but were completely unfamiliar with their life-sustaining electrical system. That second pilot on that first night was among the 10 percent of pilots who ended the session with an uncontrolled collision with the ground. Only 28 percent of the pilot participants landed safely without incident. That is correct. Not even one-third of these pilots were able to solve the problems and make the decisions that ensured the safety of the flight. All the rest had something awful happen.

Now, this was a small sample of pilots. I have no mathematical proof that these pilots were the true representatives of the total pilot population. Just because 10 percent of these pilots crashed does not mean that 10 percent of all pilots in the same situation will crash. Maybe a different group or a larger group would have done much better. But my impression as a flight instructor was that these pilots were a representative group. These were not a collection of people who did not care about their flying safely. After all, these pilots volunteered to receive extra training that was not strictly required. I may have been preaching to the choir all along. If anything, I believe that the results taken from a wider group would be worse, not better. No matter what a statistician might believe, I know that the results were scary. I believe that the problem is bigger than we once thought, and we really have our work cut out for us.

Decision flow charts

When the first simulator session was over, I asked the pilots to do me one more favor. Immediately after leaving the simulator they were seated at a desk and left alone in a room so that they could think through what had just happened and how they had reacted. I gave each pilot a sheet of paper that had this story on it:

A fire chief and a team of fire fighters arrive at a house that has smoke pouring from the rear. The chief decides to send two fire fighters into the house in an attempt to rescue anyone inside. The chief sends the remainder of the team to the side of the house to start preparing to put out the fire. With two members of the team in the house, the effort to attach hoses and prepare to spray down the house is delayed. The rear of the house then erupts into flames. The fire fighters exit the front of the house with an elderly man. By the time the rescuers rejoin the team the house is fully engulfed in flames. The chief wonders, but will never know for sure, if he had sent the entire team to fight the fire first, could they have stopped the fire from beginning and thereby saving both the elderly man and the house?

At the bottom of the page, below the story of the fire chief, was a flow chart (Fig. A-7). Then the instructions told the pilots to use the back of this page and draw their own flow chart of the decisions that they made in the flight simulator that day.

The story of the fire chief was, of course, an analogy. The fire chief was placed into a time-crunch situation where decisions had to be made. When the fire chief made the decision to split up his fire fighters, it changed his options. With the team split up, the hoses could not be prepared as quickly, but they might save people in the house faster. The fire chief could not save people and quickly fight the fire. Every decision made had an impact on the choices that pre-

The Fire Chief analogy "Flow Chart"

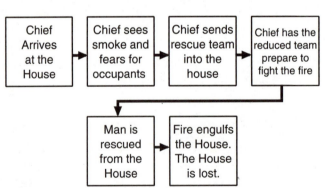

Fig. A-7

sented themselves downstream. The flight scenario that the pilots had just undergone was in many ways like the fire chief story. The pilots had each been faced with a time-crunch situation in which decisions had to be made. As the decisions were made, options changed. When pilots chose one course of action, they often could not go back and change their minds with a different course of action because time marched on and that same choice did not present itself again. I wanted to see if I could see inside the pilots' heads by using flow charts. What were they thinking when these decisions, or lack of decisions, were being made?

Some of the pilots simply did not make the analogy connection between their experience in the simulator and the fire chief story. Some did not draw a flow chart of their simulator session, but most did and I gained much insight from it.

Sample flow charts from the pilots

Figure A-8 is the flow chart of a pilot who made a safe landing. This pilot drew out the chart with all the elements included. The alternator failure is there, the loss of the glide slope, and the fact that the clouds are too low for a localizer-only approach is also there. This pilot understood all the factors that were present in the situation. Then this pilot made a clear decision about what to do after a landing at

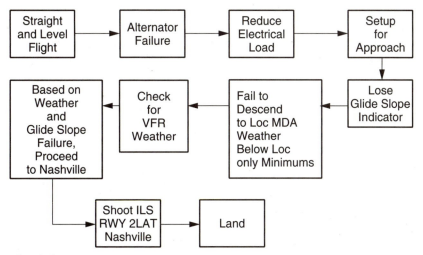

Fig. A-8

the destination was not possible. The pilot first checked for VFR weather in hopes of widening the options but then said, "Based on weather & glide slope failure, proceed to Nashville." This course of action was taken only after all the incoming bits of information had been weighed. Because all the facts had been considered, there seemed to be no hesitation with this decision. This pilot had reduced the electrical load when the alternator problem first took place, so there was plenty of battery power to shoot an ILS approach and land at Nashville. Later in the project this pilot would be classified into the information managers group (see Chap. 6).

Figure A-9 is the flow chart of a pilot who also made a safe landing, but her flight was much more hectic. The pilot recognizes the alternator failure but is unsure what to do. The pilot says, "Does not know what to do. Denial begins." The pilot was aware of the problem that the failed alternator presents but was not certain what corrective action to take and so it was easier to deny or ignore the problem. The flight environment is in motion of course. The pilot only has a few minutes to figure out this problem before something else happens and attention is diverted. The pilot says, "Considers reducing lights & radios....Fails to do so, then forgets altogether." As the instrument approach begins, the failure of the glide slope is detected, but again the pilot is not sure what to do about it. The pilot thinks for a moment that she herself might be the source of the failure. Maybe she had not turned something on correctly. This pilot said, "Glide slope goes out...unsure if failure is instrument, airport, or pilot." All these problems are putting a strain on the pilot's capacity to both fly and think. This pilot had flown the simulator well until all these failures presented themselves. "Altitude up and down like roller coaster," admits the pilot during the instrument approach.

The pilot made the missed approach after "cursing" and then the pilot's uncertainty and "I'm not sure what to do" attitude really affected the next decision. The pilot first decided to try that same approach a second time. The pilot asked for and received vectors back to that approach, but at the beginning of the approach the pilot changed her mind and asked to go to Nashville. An immediate vector was given, a Nashville approach flown, and a landing made just as the battery died. The pilot wrote, "Survive. Kiss ground....Kiss alternator." This pilot during the flight did understand what was taking place around her, but was so unsure of herself that deliberate action never took place. The decisions that were made were second-guessed and of-

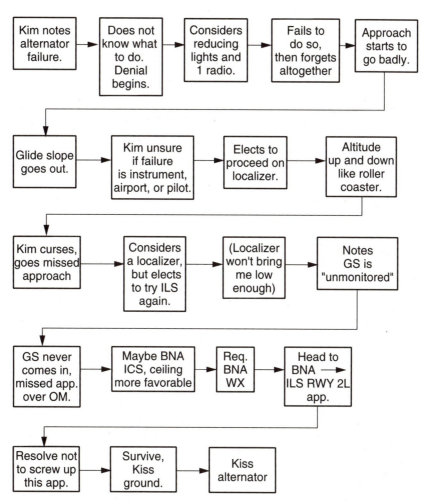

Fig. A-9 *Project pilot's Decision Flow Chart.*

ten changed. This pilot later was classified as a nonassertive decision maker (refer to Chap. 6).

Figure A-10 is the flow chart drawn by another pilot. This pilot ultimately had an in-flight electrical failure and did not land safely. The first thing that becomes obvious by looking at this pilot's chart is that not all the elements of the situation have been realized. This pilot mentions the "Alt failure," but there is no mention of the glide slope failure or the height of the clouds. The pilot at one point writes, "Shoot ILS" but in fact a full ILS was not available due to the glide slope failure. After the first attempt at the approach ends with a "Missed Appr," the pilot asks for a second approach. There has been

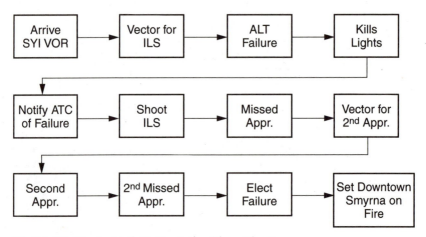

Fig. A-10 *Project pilot's Decision Flow Chart.*

no recognition that the clouds are lower than the airplane can fly on the upcoming approach, so the second approach as well ends with "2nd missed appr." Time runs out. The pilot writes, "Elect failure....Set downtown Smyrna on fire."

Remember that this pilot had been given the exact same scenario as all the others, but this pilot just did not see and therefore could not use vital information for decision making. This pilot got behind the airplane early and was simply swept away. The pilot made attempts to get things under control, but anytime progress was made, three other tasks would present themselves and the pilot would be behind on them. The pilot's lack of problem recognition and reaction began to feed itself. Like a snowball rolling down hill, the magnitude of the pilot's problems began to grow larger and larger. There came a point of no return when it was obvious that this pilot would not and could not pull himself together in time. I saw this crash coming long before impact. I called pilots who got caught up in this way the "snowball effect" and that became one of the pilot categories identified later in the project.

Figure A-11 is the chart drawn by a pilot who also experienced an in-flight electrical failure and a probable airplane accident. This pilot never mentions the alternator failure or the cloud heights anywhere on the event flow chart. The pilot does ask the question in one box, "GS out?" but there is no evidence that the impact of a failed glide slope is realized by the pilot. The pilot writes, "Missed approach too late." In fact this pilot flew past the missed-approach point by *6 miles* before beginning the missed-approach procedure. Pilots take note: There is zero obstruction clearance provided by the minimum descent altitude past the missed-approach point. If you do not climb out at the proper position, the chances of flying into terrain, towers, buildings, and so forth, are great. This pilot, expecting to see the runway, flew on and on at the MDA rather than executing the missed approach. The pilot made no decision at all and probably would have had an accident right then and there. Ultimately, the pilot did start a missed approach and asked for a second try at the exact same approach. The electrical system failed on this pilot during that second approach attempt.

The pilots who drew the flowcharts in Figs. A-9 and A-10 had wild rides. They had a hectic, frustrating, sweat-inducing roller coaster of a flight. The pilot who drew the Fig. A-11 flow chart acted calm the whole time. It was as if this pilot had nothing to worry about and was oblivious to all the mayhem that was circling around her. On a scale of 1 to 3, where 3 represents total situation awareness, and 1 represents a total lack of awareness, this pilot was not even on the scale. This pilot was just unaware that anything was happening that should be of any concern. I could not help but picture this pilot strolling across a battle-field between enemy foxholes. As bullets whiz past from every direc-tion, the pilot is thinking, What a nice day this is for a walk.

Figure A-12 is the flow chart of another pilot who was in the sim-ulator physically but did not seem to be there mentally. He writes in two places that there is a "simulator malfunction." The first prob-

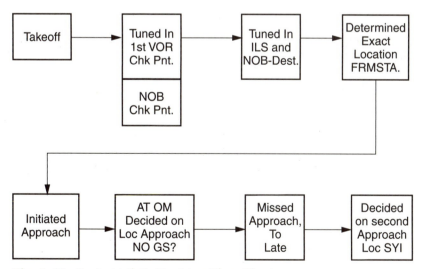

Fig. A-11 *Project pilot's Decision Flow Chart.*

lem that this pilot perceived as a simulator malfunction was in fact the pilot selecting an incorrect frequency for the localizer. Of course this produced an OFF indication that the pilot assumed was the simulator's fault. I never understood where the incorrect frequency had even came from because this pilot never bothered to look at any approach charts. The second "malfunction" was the middle marker light and sound coming on when the pilot believed he was passing the outer marker. This pilot ends the flow chart with the notation, "Missed Approach." In fact, this pilot began the descent to an unknown MDA at the middle marker, flew past the airport by 8.4 miles, completed a missed approach, asked for the same approach again, and had electric failure during the vector for the second approach.

The pilots of Figs. A-11 and A-12 never got worried. Later I called these pilots the lost in space group. I know that is not a very delicate and certainly not a very scientific name, but they could be named nothing else.

Airliners have the cockpit voice and flight data recorders, called the *black boxes* (even though they are actually orange) to record the conversations and actions of pilots. These recordings are used to get some idea what the pilots are thinking. After an accident, investigators listen to the words of the pilots in an attempt to understand what the pilots were thinking and what was driving their

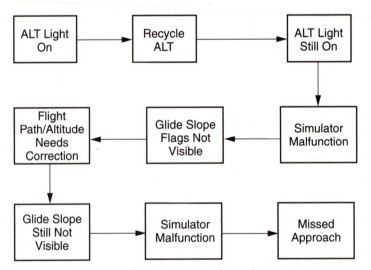

Fig. A-12 *Project pilot's Decision Flow Chart.*

decisions. Even with this equipment it is not possible to truly know what the pilot's inner thoughts were. But, words spoken come from the thoughts of the mind, so hearing the words is a window to the mind.

In general aviation we have no black boxes. I guess it will be impossible to ever know exactly what is running through a pilot's mind as critical decisions are being made, but the flow chart was a stab at it. Studying the flow charts I began to see patterns. I saw that when the pilots were able to take in information and use that information in their decision making, their flight outcomes were most often favorable. When pilots missed information that was presented, their decisions were flawed and their outcomes unfavorable.

Getting inside the mind is tough. I'm sure there were many facets of the pilot decision-making process that I did not see or saw but did not understand. But, as a pilot, it was fascinating to watch. Do not ever let anyone tell you that piloting is a physical task, a trade that can be mastered, a technical skill that can be copied like blueprints. Piloting is a complex mental process that is different each time you fly and is different for each individual who flies.

The expert pilots tested

When the original pool of general aviation pilots was asked to participate in the decision-making project, experts were eliminated. I wanted to see the performance of average GA pilots, not pilots who flew every day for a living. But the more problems that the GA pilots had, the more interested I became in what the expert's performance would be. So while the GA pilots were flying the first LOFT scenario, I invited other pilots who would be considered experts to come and give the simulator a try. I defined an expert for the purposes of the study as a person who regularly flew as a normal job function. The experts I invited were FAA inspectors, corporation pilots, Part 135 on-demand charter pilots, scheduled air cargo pilots, and active instrument flight instructors. These professionals were given the same first LOFT scenario as the larger group of GA pilots. The experts were given the same instructions, the same ATC procedures, and the same exact set of problems. Did the experts do a better job in the scenario? The difference between the GA pilots and the experts was profound.

The expert pilots flew the same scenario that was "killing" the GA pilots, but every expert landed safely. As with the GA pilots, a timer was kept on the experts during their flight scenario. The expert's time interval between the onset of the problems and resolution of the problems varied to some degree, but they ultimately all made the same decision. One hundred percent of the expert pilots landed without incident after following essentially the same decision sequence. After observing the same performance again and again, I became convinced that I was watching expert flight behavior.

I was very concerned about the performance of the experts before they began. If the experts all attacked the problem differently, it would seem to indicate that flight performance was very individualistic. It would mean that emulating expert behavior would have been impossible because there would have been as many behaviors as there were experts. But that did not happen. The experts all followed what could only be termed an expert logic, and every one of them was clearly the pilot in command.

The expert's flow charts

Figure A-13 is the flow chart drawn by one of the expert pilots after she flew the first simulator scenario. This chart was typical of the charts drawn by the experts. You can actually see the logic flow through the diagram. The expert catches all the incoming information and processes it almost immediately. The expert who drew the Fig. A-13 chart writes, "Alt failure—shut off equipment—use only comm/nav 1 and ADF." Not only did this expert notice the problem, but she also understood the gravity of the problem and quickly shut off all nonessential equipment. Then the expert made a list of the equipment that is considered essential to get back on the ground. The expert flew the first approach to minimum altitude, did not see the runway, executed a missed approach, and then without delay made a decision: "Call BNA (Nashville) Request ILS 2R." While en route to BNA and knowing that the battery was depleting, the expert asked for and received a turn at the marker. This maneuver meant that the pilot would have to make an extremely tight localizer intercept, but the expert was willing to take and anticipate that quick turn rather than waste more time on a long vector to the localizer. The chart concludes with "Shoot the App and Ldg." Remember, every pilot that was invited to the project as an expert landed safely.

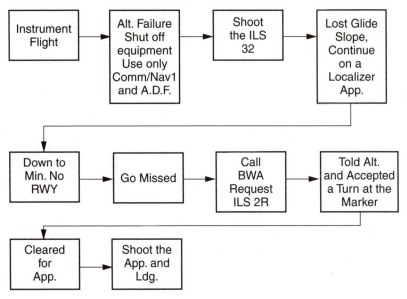

Fig. A-13 *Project pilot's Decision Flow Chart.*

The expert scale

It was clear that expert and novice pilots were different, but how different? Knowing just how close or how far apart these two groups were would help in understanding how big a job it would be to raise the novice pilot performance up to the level of the expert. To help answer this question I designed an expert scale. Only the first simulator session performance was used for this comparison because this session was flown by both the experts and novice volunteers. I constructed the expert scale by using a calculation that statisticians would call a transformed Z score. The expert scale was calculated by transforming the average expert decision time into the one hundredth percentile on a scale from 0 to 100. The novice pilot's performance was transformed in the same fashion and placed into a percentile rank relative to the expert's performance of 100. As you would expect, some of the novice pilots did better than others. Some even came close to equaling the standard set by the experts, but on average the novice pilots took twice as much time to come to a decision as the experts. The average ranking on the expert scale of all the novice pilots taken together was 51.835 percent. That means that the participants as a whole ranked approximately at the fifty-second percentile position in comparison with the expert performance. This means that on average a novice instrument pilot is only about half as fast in making decisions as experts. Described another way, experts make decisions approximately twice as fast as novices. And remember these experts were not just making fast decisions, they were making correct decisions. Every expert pilot landed safely because of his or her quick and accurate analysis of the situation. In the meantime, the novice pilots were taking twice as long to figure out what to do, and when they did decide, their decision was not always accurate. Also the nature of the scenario called for a quick response to the events taking place. Remember, the alternator had failed and therefore electrical power was fleeting. When time was critical, the novice pilots "wasted" life-saving time as they struggled with a decision or, worse yet, were not aware that a decision was called for.

The general aviation pilots divided into groups

After all the pilot volunteers had flown the flight simulator for the first time, I invited them back for a seminar. Up to this point in the project all the volunteers had the same experience with the project. They all had scheduled a time to come and fly the first scenario. They all had the same electrical failure, they had filled out all the same paperwork, and all had drawn the flow chart following the session. But what the volunteers did not know was that they had been randomly divided into two groups. The seminars were held at the university where I teach, and there were plenty of classrooms that could have held the entire group, but I deliberately used a smaller classroom. This way, when I divided the two groups, I could tell the volunteers that the reason there was going to be two groups and two seminars was because the room would not hold the entire group at once. This was just a gimmick so that the volunteers would believe that both seminars were alike. But in fact they were not anything alike.

In the morning session, the volunteers arrived and I taught a seminar on basic instrument flight. I answered questions about flight instrument systems and IFR clearances and ATC procedures. It was informative but generic. We did discuss the events from the first flight scenario. We talked about the electric system, how it worked, and what that red light really meant. All in all that seminar was a traditional lecture on instrument flight. The volunteers in that morning session became the traditional group, and they left the seminar thinking that the afternoon session would be the same as what they had just attended. In reality the morning group had been given a placebo.

It is important to remember that these pilots were self-selected volunteers from the Nashville area. I was concerned that pilots who knew each other might be in opposite groups and that they might discuss not only the seminars but also the flight scenarios themselves. If they had discussed it, they could have realized that the seminars were different, and they could have known what was coming in the simulator. In the project paperwork there was a request that participants signed that asked them not to discuss the events of the project with anyone until after the project. I did not believe that that alone would prevent some cross-pollination but as it turned out it was not an issue. These pilots really did not know each other outside the project. You would think that entire pilots' groups [Experimental Aircraft Association (EAA) chapters, Civil Air Patrol (CAP), etc.] would have participated as a block, but that did not happen. I was aware of only two pilots in the project who were friends and had flown together prior to the project. They both scheduled their first simulator session on the same night, and I made sure that they actually were in the simulators at the same time so one could not tip off the other. Those two pilots were also in the same seminar group as well. The entire project was conducted over the shortest possible stretch of time to further reduce the chance of participant discussions. There may have been some breeches of the project protocol, but I was convinced that if it did happen, it was isolated and did not affect the outcomes.

When the afternoon group of volunteers arrived for their seminar, I had reconfigured the classroom. The afternoon group would receive a completely different seminar, this one targeting pilot decision making, not simply instrument flight. I wanted to give the second group an understanding of naturalistic decision theory and how it applies to the pilot. These pilots became known as the naturalistic group.

I laid out the basic concepts of naturalistic decision making: decision making under stress, time pressures, and high stakes. Then I introduced the ASAP model. Each participant was given a card (Fig. A-14). The card served as a quick reminder of what the letters ASAP stood for in this example. Most people understand the letters ASAP to mean "as soon as possible," and that metaphor of urgency applies to pilots who must make decisions under stress, but the letters ASAP

```
┌─────────────────────────────────┐
│                                 │
│     ASAP DECISION MODEL         │
│       Anticipation              │
│       Situation Awareness       │
│          Action                 │
│           Preparation           │
│                                 │
└─────────────────────────────────┘
```

Fig. A-14

for the naturalistic group of pilots stood for anticipate, situation awareness, action, and preparation (see Chap. 5). ASAP became just a memory jog. It was something to help them concentrate and zero in on the problem so they could more easily solve the problem.

Using associated groups for problem solving

When the volunteer pilots flew the first simulator scenario, they were presented with a problem. They had a red alternator warning light illuminate while setting up for an instrument approach. All the pilots saw the light, but most did not react to its warning in any way. Later some of the pilots said that they knew the light represented a problem, but they just did not have the time to think it through to a solution. The whole point of using associated groups is to zero in on the problem without using a lot of time, because anything that did require a lot of time would not be practical or usable during this time-crunch situation.

Figure A-15 illustrates the problem that confronted the pilots in their first trip in the simulator. Everything was working well and all was right with the world, and then the alternator fail light came on. The pilots were then faced with a wide array of instruments, switches, dials, and levers. What should be done now? Which of all these instruments and switches should I be doing something with? Will I even have time, while flying the airplane, to think about it? The pilots faced a huge list of possible associations. Some of these are listed in the second box of Fig. A-15. These possible associations are all mixed up, there is no order, and the items on the list are not grouped into families. Most of the pilots never got past this point. They had an information overload. There were just too many possibilities, and there was just not enough time to sort them all out. But some pilots, and all the expert pilots, were able to identify the problem. These pilots used only the instrument, gauges, and switches that were from the group that would have contained the alternator fail light. The third box of Fig. A-15 removes the items from the second box that belong in the alternator light's associated group. The vacuum pump, magnetic compass, and suction gauge have nothing

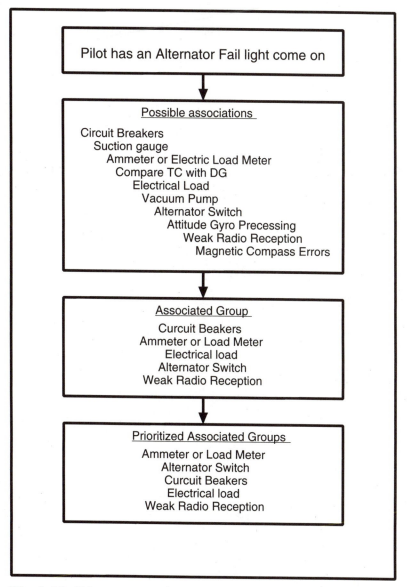

Fig. A-15 *Using associated groups for decision making.*

to do with the alternator system, so they were eliminated because they are not in the family. But the circuit breakers, ammeter, electrical load on the system, the alternator side of the master switch, and the possibility of weak radio reception all do go together. These are all associated in some way. The pilots who used this associated group now had a narrowed list of items to troubleshoot.

The pilots who really seemed to know the system took the associated group and prioritized the members of the group. Box 4 of Fig. A-15 indicates how these pilots prioritized the list. They first checked the ammeter for a discharge of the battery that would have verified that the problem was in the electrical system and that the culprit was probably the alternator. Next they tried to reactivate the alternator by cycling the alternator switch off and back on. This could create a better connection to electric flow to the alternator, which might restore its function. In the scenario the pilots first flew, cycling the alternator switch did not restore the alternator. This meant that the alternator had truly failed and that for the rest of the flight all electrical power must come from the battery. The magnitude of this problem did not sink in with most of the pilots. When flying in the clouds, the only safe way back down to Earth is to fly an instrument approach. You cannot fly an instrument approach without some navigation radio, which use electricity. If your battery runs out of electricity, you will not be able to operate the radios or fly the approach or get back down safely. The alternator light became a life and death situation. The associated group led the pilots to this critical understanding. They must save as much electrical power as possible so that there will be enough power to fly the approach to safety. Here is where a few pilots reduced the electrical load to save the power for the approach. It made no sense to flash strobe lights, burn landing lights, and power radios that were not needed—but many pilots left these items on, and their chances of a safe outcome started to die with their battery. As their electrical power went away, their interior lights would start to dim and their radio reception would start to weaken and crackle. Eventually all the electricity would be gone, and they would be flying around lost in the dark with no safe way down. A very scary situation.

Sixty percent of the pilots who flew the alternator-light scenario did not troubleshoot the electrical problem in any way. In other words, 60 percent of the pilots were roadblocked by either a lack of time or a lack of alternator understanding or both. Some knew all about the electrical problem and what to do about it, but they had no mental automatic pilot (see Chap. 5). They could not both fly the simulator and work on the problem. Others flew the simulator well but had no idea how to link the alternator light to its associated group. They had some time to work the problem, but they did not know what to do or where to begin. And some pilots had neither the time nor the knowledge. They were literally shooting in the dark. Forty percent of the pilots experienced the electrical failure in flight. This means that they lost the race. The battery died before they could get on the ground.

For that 40 percent of pilots the possibility of a survival outcome was slim. They would have had to descend through the clouds on an unknown course because they would have had no working VOR, ADF, GPS, or anything. They could have been near high terrain or towers or buildings and never known it as they descended. If they came out from under the clouds with enough room to maneuver, they would have been fortunate to see anything flat enough to land on: a highway or a field. But the scenario was at night so that would have made spotting a landing site even less likely. Surely, many if not most of these pilots would have had a fatal accident if this situation had actually occurred in flight. After this project I think every one of those volunteer pilots ran out to buy a hand-held radio and now never go anywhere without a fully charged cellular phone.

The pilots who did have time to work on the problem and did troubleshoot with the aid of associated groups bought themselves some time. The battery was going to die no matter what, but by understanding the system, they knew that reducing the electrical load would extend their time and their possibilities. Every expert who flew the scenario made a pattern match. The experts knew that the alternator light meant a draining battery, and they matched a solution to the problem. They used associated groups to verify the problem and reduced the effects of the problem. All the experts addressed the correct problem and all landed safely.

Using single-pilot CRM

One of the most amazing examples of single-pilot CRM that I ever saw took place one night in our flight simulator during the volunteer pilots' first scenario. While turning onto the localizer with a glide slope inoperative flag showing, one of the participant pilots reached down into her purse, which she had taken with her into the simulator. She pulled out her cellular phone. She flew the simulator with one hand and dialed up her flight instructor with the other. Her flight instructor actually answered the phone and she asked about a localizer approach when the glide slope was out. She got her answer, returned the phone to her purse, and correctly flew the approach.

The second LOFT session

After the volunteers had flown the first LOFT scenario and attended the workshop/seminars, they were scheduled back into the flight simulators for a second session. The second LOFT scenario was different from the first, but it featured the same elements, including decision-prompting situations. The results were overwhelmingly positive. Pilot volunteers from both groups made statistically significant improvement in decision times, decision quality, and scenario outcomes. In addition, pilots who had been given the specialized decision training had 10 percent fewer accidents than did the group trained with traditional methods

The second scenario began at the Perry County Airport in Linden, Tennessee (Figure A-16 position A) with the intended destination airport of Nashville International Airport (Fig. A-16 position B). The participants were told that the purpose of the flight was to deliver a shipment of blood for the American Red Cross. An Igloo cooler with the sign "BLOOD DONATION—American Red Cross—Rush Shipment to Nashville—Highest Priority" was placed in the cockpit of the simulator with the participant (Fig. A-17). The participants were also told that the airplane's radio call sign for this particular flight would be changed from Frasca 141 to Lifeguard 141 because the blood donation was on board. In the second scenario I wanted the pilots to have a good reason to be taking this flight. The blood donation provided a purpose and also a sense of urgency. Many of the pilots had treated the first scenario as a training exercise and did not really and truly believe it to be real. By placing the blood donation container in the cockpit with the pilots, I wanted to give them a mission to accomplish.

As with the first session, a set of written instructions was given to each pilot upon entering the simulator. Figure A-18 is a copy of the instructions given for the second session.

After takeoff the route of flight took the airplane to the Graham VOR station near Centerville, Tennessee. As the flight passed over the Graham VOR station, I began the timer. As the flight neared Nashville, this weather report was given to the participants:

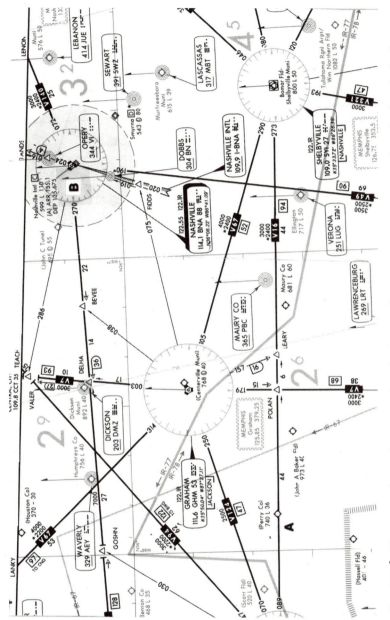

Fig. A-16 *The second simulator scenario began at the Perry County Airport (marked with the letter A in the lower left of the chart) to Nashville International Airport (marked with the letter B in the upper right of the chart).*

219

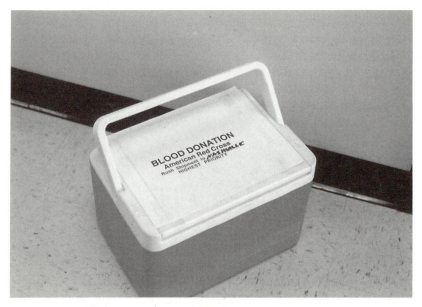

Fig. A-17 *This ice chest was placed inside the simulator during the second simulator session to give the pilots a "mission" to accomplish.*

"Nashville has a measured ceiling of 600 overcast, visibility is 1¹/2 miles, wind 360° and 10 knots, radar indicates a level 3 thunderstorm that is 15 miles northwest of the Nashville Airport, moving southeast at 12 knots." The participants could have understood from this report that the cloud ceiling was high enough for any approach at Nashville and that even though there was a thunderstorm moving toward the Nashville Airport, it would not reach the airport for over 1 hour.

Figure A-19 is a photo of the engine instruments when they are indicating normal operations. These instruments are just in front of the control wheel on the instrument panel. The oil pressure, oil temperature, and cylinder head temperature are all indicating in the green arc. During the weather report, I set the simulator's computer to slowly indicate the onset of a decrease in engine oil pressure and a simultaneous rise in engine temperatures. Figure A-20 is a photo of the engine instruments after the oil loss and temperature rise had taken place. All three now indicated on the red lines. These indications tell the pilot that the engine is losing its oil and that engine failure is imminent and unavoidable. These indications should have told the pilot that the remainder of the scenario was a race against time before engine failure.

Instructions before entering the Second
simulator session.

1. As in the first session, you will be given a few minutes at the beginning of the session to fly the simulator. As you now know the simulator will not "fly" exactly like an airplane. The simulator is for "procedures" training not necessarily "flight skill" training.

2. You will start the second simulator session in Linden, Tennessee at the Perry County Airport. Your destination is Nashville International Airport (BNA). After some "warm-up" you will fly to the Graham VOR at 4,000 feet. You will be flying in the clouds and on an IFR flight plan.

3. As before, you will need to wear a headset that is provided (or you can use your own). All conversation that takes place between you and the instructor after he says "lets go" must be as conversations between a pilot and an air traffic controller. The instructor is not allowed to give "instruction" but only functions as your controller. You should feel free to make requests of the instructor just as you would an air traffic controller. In actual flight, when frequency changes are made a new voice would be heard. In this simulation, all frequency changes (Memphis Center, Nashville Approach, Smyrna Tower, Nashville ATIS, Nashville Tower) will take place as normal but it will be the same voice. You are to ignore this fact and assume that it is a new controller after each frequency change.

4. You will be given a current set of approach charts (you may choose either NOS or Jeppessen or use your own). Once you begin, you may assume that all frequencies listed in the approach charts are operating properly unless otherwise noted or indicated by OFF flags. This will include ATIS, AWOS, Control Towers, Approach Controls, Center Controllers, VORs, NDBs, Marker Beacons, Localizers & Glide Slopes, etc.

5. The second simulator session will also use the cockpit light. Anytime that you fly the simulator to a position that is both below the clouds and in sight of a runway the light will come on. The instructor will demonstrate this light before you begin.

6. **Please treat this simulation, in every way, as if you were actually in flight and situations presented are actually happening to you!**

7. Also as before we ask that you not discuss the second simulator session with any other pilots at least until November 8, 1997.

8. The weather at Nashville is 600 Overcast. All other airports have ceiling lower than BNA. There are no pilot reports or NOTAMs that are significant to your flight. All your instruments, avionics, and other equipment are operational prior to takeoff.

9. Good Luck! Have some fun, and once again thank you for participating.

Fig. A-18

The pilots were then told by the air traffic controller to "Expect the ILS approach runway 2 Left at Nashville" (Fig. A-21). But shortly after these instructions were given, the pilots were called by the controller and given this message: "Lifeguard 141, there has been an airplane land at Nashville on runway 31 with the landing gear up. Rescue crews are now on the way to the scene of the accident.

Fig. A-19 *Engine instruments reading normal in the green arcs.*

Until the rescue crews tell us that everything is all clear, the Nashville Airport is temporarily closed. You can still expect the ILS to runway 2 left, but you will be getting an extended vector for this delay."

Within 1 minute of this announcement, the airplane's engine began to run rough due to the loss of oil in the engine. I used the simulator's computer terminal to cause the engine noise to hesitate and the tachometer to show rough and wavering rpms. This simulated the

Fig. A-20 *Engine instruments reading in the red arcs.*

rough-running engine. The pilots, without their knowledge, were given 17 minutes from the onset of oil loss indications until engine failure. Unless the pilot objected, the controller then gave the pilot a turn away from three airports (Nashville's International Airport, Nashville's John C. Tune Airport, and Smyrna's Airport) to delay the approach to Nashville because of the accident at the Nashville airport. I recorded time intervals and made observations of the decisions the participants made or failed to make past this point of decision.

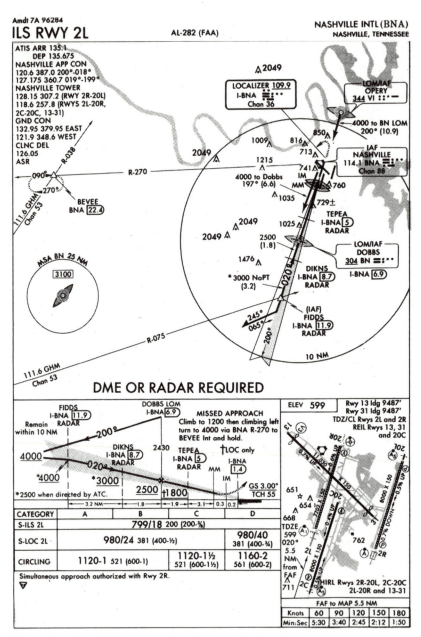

Fig. A-21 *The ILS Runway 2 Left at Nashville, Tennessee.*

Some of the pilot participants rejected the turn and requested a diversion to another airport. Some of the pilots said nothing about the engine problem to the controller and accepted the delaying turn. If the pilot accepted the turn, the controller vectored them but eventually told them to proceed to the Fidds Intersection and hold along the localizer. There were a small number of pilots who actually accepted this hold, despite the engine's continued faltering, and ultimately (when the 17 minutes ran out) experienced an engine failure while in the hold. These pilots never mentioned their engine problem to the controller. Of the pilots who elected to divert to another airport, some landed safely and some did not. It depended entirely on which airport they selected and how well they flew to get there. One of the area airports was the John C. Tune airport, which was in the approximate location of the reported thunderstorm. At least one pilot selected the Tune airport despite the thunderstorm report. When the simulator-generated turbulence got too much for that pilot, he asked to turn around. The controller granted the request, but the pilot had wasted too much time flying into a thunderstorm and his engine failed en route to this second choice. There was another small group of pilots who, when faced with the delaying turn, simply declared an emergency, flew the ILS runway 2L approach at Nashville, and landed safely with plenty of time remaining. This solution was the quickest and easiest way to get on the ground, yet most of the pilots were reluctant to declare an emergency. This reluctance became a great roadblock to safety and is discussed in greater detail in Chap. 7.

During the second scenario 19 of 58 pilots, or approximately 33 percent, experienced an engine failure in flight and in the clouds. These pilots either misinterpreted the engine problem as "not too bad" or just could not come up with a plan fast enough to get on the ground prior to engine failure. That is the bad news. The good news is that the other 67 percent (39 of the 58) of pilots landed safety and without incident. Recall that during the first scenario only 28 percent landed safely without incident. This was very encouraging. Of course, during the second scenario, the pilots knew more of what to expect, and maybe that is why they did so much better. But isn't that the whole point. If we can get pilots to expect to make decisions and take action on those decisions, we will be safer.

What about the performance of the two groups? The pilots who were in the naturalistic group and received the decision training, including the ASAP model, landed safely on 70 percent of their flights.

The traditional group, who had not received the specialized decision training, landed safely on 61 percent of their flights. The difference between 70 and 61 is not statistically significant, but it is still important. It appeared that the decision training that used naturalistic decision making techniques had made some, albeit small, impact. But remember, the pilots who experienced an engine failure while flying in the clouds faced a very uncertain and dangerous immediate future. I could not say with certainty that they would have been killed had this actually happened in an airplane, but the chances of survival or escape without injury would have been slim. Because we are dealing with life and death, I was happy with any improvement, statistically significant or not.

The bottom line was that all these pilots received additional training that they would not have received without the project, and both groups improved. So I guess this tells us what we knew all along: Training is a good thing. But more was learned. It was not just any training, but training with the use of real-world flight scenarios that made significant improvements in a pilot's ability to deal with problems and produce favorable outcomes. The pilot participants of this project were previously unexposed to the techniques used in this project. These were civilian pilots who were trained by learning maneuvers and meeting FAA mandated training requirements. But the procedures used in this project were taken from the airline, corporation, and military flight training environments. The data analysis produced several important discoveries that were covered in earlier chapters, but none has greater impact than the fact that civilian, low-experience pilots responded with such great improvements using techniques that were "stolen" from the airlines and the military.

The numbers

When all the pilots who volunteered for the project had gone home, when all the simulator sessions were over, when the seminars had concluded, I found myself alone with a mountain of data, numbers, observations, journals, you name it. The big job then was to try to make some sense out of all that I had seen. In an attempt to understand all this information I divided my attack into two paths. First, I used a quantitative approach to see if the information proved anything. For this job I used a series of mathematical tests. The results of those tests and an explanation of their results, appear later in this appendix. Second, I used a qualitative approach, the results of which are contained in the previous chapter.

I know most pilots hate statistics, and you may even find yourself skipping this section but give it a try anyway because some of the information was quite intriguing.

The pilots

It would not be possible (in my lifetime) to have every single general aviation pilot in the United States participate in a decision-making project such as the one I undertook. Because not every pilot could participate, I hoped that the pilots who did would represent all the other pilots well. In other words, I hoped that the talents and short-comings displayed by the group of pilots in my project would exactly mirror the talents and shortcomings of the entire pilot population. In the end I cannot say for sure that they did, but there are some things that I can say.

The participants began their first flight training device sessions on September 13, 1997. At this point the random selection of participants to a group had taken place, but during the first sessions none of the participants knew that groups had been selected. On October 9, 1997, 65 sessions had been completed. It became important to certify that the randomly selected groups performed with relative equality. One group could not have been made up of more experienced or more expert members than the other. If later I was going to be able to make comparisons between the groups, I needed to know from the beginning that both groups were starting off from the same point. I had divided the pilots up randomly, but I still had to make sure that one group was not accidentally made up of a bunch of "ringers."

To test this first question the decision times of the participants dur-ing the first simulator session were calculated. The decision times were left in seconds rather than minutes and seconds for easier cal-culation. The fact that a unit of time has an absolute zero, or starting point, and that every unit of time is equivalent to every other unit makes the "seconds data" the ratio scale type. This is important be-cause if data can be considered ratio scale data, several statistical tests are appropriate for use in evaluation.

To make sure that the two groups were equal representatives of the pilot population, I used an independent samples test. This mathematical test would prove whether or not the decision times of the two groups were about the same or if the groups were different to a statistically significant degree. When this statistical test was calculated, the two pilot groups came out about even (t (56) = 0.568, with 2.021 critical at $p < .05$). The two samples were made up of pilots who had displayed equal decision-making skills in the first simulator sessions. This information was extremely important. It proved that the random selection to groups had been successful and that no group was superior to the other before any treatment began.

The use of experts was also very important to the results of the project. As mentioned in Chap. 4, to qualify as an expert for the purposes of this project a pilot must have been a current instrument pilot and employed in a capacity where instrument flight can be encountered on a daily basis. It became important to show that the experts who were used in the study were in fact from a different population of pilots than the population that the participants were drawn from.

The concepts of expert and pilot in command are somewhat abstract, and I did not want to base any conclusions on just what I thought an expert was or what I thought a pilot in command should be. So I conducted more statistical tests to see if in fact these pilots were different from the population of average general aviation pilots. To do this I compared the decision times of the volunteer pilots (from both groups) with the decision times of the experts. This test revealed a statistically significant difference between the volunteer and expert pilots of the project. In other words, comparing the project's pilots to the expert pilots was like comparing apples and oranges.

The results of the comparison between the experts and the traditional group yielded a significant difference (t (34) = 3.146, with 2.042 critical at $p < .05$). Likewise the comparison of the experts and the naturalistic group produced a significant difference (t (38) = 2.825, with 2.042 critical at $p < .05$). From these two tests there is evidence that suggested that the experts were in fact experts and that they had been drawn from a different population of pilots than were the project participants.

The first set of simulator sessions established a baseline of performance for the participants. It also showed that the participant groups were samples from the same population and also proved that the experts were not from the same population as the participants. These initial tests did not indicate any benefits from the project but produced a starting point. The initial tests acted to qualify the groups and the experts for the remainder of the project. The pilots returned after their first simulator session for a seminar. The pilot groups met separately, but members in a group did not know that the seminars were different.

At the end of the seminars the group members were asked to sign up for their second session in the simulator. Two different sign-up sheets were used. The two sign-up sheets staggered the sessions by date and time. This was to prevent the first group from monopolizing the early sessions and thereby pushing back or extending the time between simulator sessions for the group that signed up last. The sign-up schedules gave equal opportunity to both groups for early sessions.

On October 13, 1997, the second simulator sessions began. The last session was completed on November 25, 1997. The most obvious question quickly became Did the group performance improve in the second session over the first session? To help answer this question the decision times in the second simulator session were tabulated and compared with the decision times of the first session. In this test an individual's performances in both sessions were compared. This comparison required the use of a dependent samples test. This type of test is often used when the same group of subjects is tested twice, as in a pretest-posttest study.

A first dependent test was conducted between the traditional group's first performance and that same group's second performance. If no significant difference is found, that would indicate that "nothing special" had taken place between the simulator sessions. If significant difference is found, that would lend support to the statement that improvements had been made in pilot decision skills. The test of the before and after skills of the traditional group did produce evidence of improvement ($t(28) = 7.075$, with 2.060 critical at $p < .05$). This test showed that there was a statistically significant improvement made by the traditional group.

A similar dependent samples test was conducted using a comparison between the naturalistic group's first and second simulator session performance. Once again, statistically significant improvements were supported by the data (t (30) = 8.866, with 2.045 critical at $p < .05$). Working with numbers you must be careful not to claim too much. The Earth did not move when these numbers came out the way they did, but I was very happy.

These tests indicated that a positive training effect took place within this project. It also appears that the naturalistic group, with an 8.866 t-statistic, made greater improvements than the traditional group with its 7.075 t-statistic.

Both groups made statistically significant improvements in their ability to make decisions under pressure and in a timely manner. The group that was exposed to naturalistic decision-making (NDM) concepts in a seminar did improve more than the group that was exposed to only traditional training concepts. However, the greater improvement by the naturalistic group over the traditional group was not statistically significant. The one, 2-hour seminar had effected the outcome some but not to a large degree. I now wonder if exposing the pilots to more NDM concepts over a longer period of time would have produced a bigger gap between the groups. Aside from that, the big news was that both groups did a lot better the second time around. The project did a good job of helping the pilots focus on the problem and showed that a little instruction can go a long way. Put another way: Training is a good thing.

Test results

Decision times alone did not tell the whole story of what was going on. There were instances where a participant made a late decision, but the simulated flight landed safely despite the participant's lack of speed. In a case such as this the decision time was poor, but the outcome of the flight was positive. In other cases the decision time was relatively fast, yet the simulated flight did not end with a safe landing.

My outcomes evaluation began by reviewing the possible decisions that were made by each participant while piloting the simulator and the outcomes of each decision. During the first simulator session five different decisions were observed. These decisions were then ranked, with most favorable indicated by a 5 and least favorable indicated by a 1. The first simulator session decisions that I saw were

1 Accidentally, or without knowing the relationship between the MDA and cloud bases, descended below clouds. If the airport was seen by the pilot, it was an accident and not the result of instrument flight technique.

2 Missed the first approach at MQY, but then elected to try another approach at MQY one or more times.

3 Knowingly and strategically descended below clouds on the proper approach path to emergency landing at MQY on the first attempt at MQY.

4 Missed the first approach at MQY and diverted to an alternate when asked their intentions.

5 Diverted to alternate without a first approach at MQY.

After the participants made their decision, one of three outcomes was observed:

1 The simulated flight progressed to a safe landing without any extraordinary situation developing. This condition was coded LS for landed safely.

2 The simulated flight progressed to a landing, but some unusual circumstances developed. An example of this was an emergency descent below a safe approach altitude through the clouds to a runway. Another example is an engine failure in flight, but the pilot was within glide distance to a runway and successfully accomplished the glide to landing. This condition was coded LU for landing at an airport under unusual circumstances.

3 The simulated flight progressed to a point where an airplane accident took place and the loss of life was possible. An example of this situation was a crash landing that did not take place at an airport, loss of airplane control, or a complete electrical failure in the clouds. This condition was coded PA for probable airplane accident.

The numbers, 1 through 5, that were used to identify the individual decisions that were made during the first simulator session do not represent differences in magnitude, as would be the case with a unit of time. The numbers selected are used simply to distinguish a category and are therefore nonparametric. This data is ordinal scale data because they were ranked on a continuum with 5 being most favorable and 1 being least favorable. Ordinal data, being nonparametric, required a new statistical test because a t-test would not be appropriate. The statistical test that is most appropriate for this data is the one-way Chi-square (pronounced ki square and coded X^2) test for qualitative data.

When using units of time and an appropriate *t*-test earlier in the evaluation of the data, it was important to demonstrate that the two groups that were to be evaluated were from the same population. That determination was made using time, which was a ratio scale data. At this point, working with ordinal scale data, the same question was asked: Did members of the two groups make statistically similar decisions and therefore can they be considered from the same population of pilots? A X^2 test was calculated to support the claim that the members of each group were of the same pilot population and that neither group started out superior to the other.

The X^2 test was begun with the following logic: If the participants from a first group were observed to perform in a certain way, and if all participants were from the same overlying population, a second group could be expected to perform as the first group did. With this logic in place the first X^2 test was conducted with the traditional group's first simulator performance as the observed frequency and the naturalistic group's first simulator performance as the expected frequency. The naturalistic group had two more members than did the traditional group, so proportions were used to calculate the expected frequencies. When this one-way X^2 test was completed, once again there was statistical evidence (X^2 (5) = 2.797, with 9.49 critical at $p < .05$) showing that the groups were the same based on the decisions made during their first simulator session.

Another one-way X^2 test was also calculated on the groups based on their outcomes. Again the outcomes were coded LS for land safely, LU for land under unusual circumstances, or PA for probable airplane accident. Using the same format and logic as the decisions X^2 test before, this outcomes X^2 test carried evidence of group similarity. The test returned a X^2 statistic that was not greater than critical (X^2 (3) = 1.299, with 5.99 critical at $p < .05$), and therefore this test gave further evidence that the two groups, starting out with the first session, were equivalent and that no group was unique.

X^2 tests showed again that the experts were indeed experts. Two X^2 tests were conducted that compared the participant group as a whole with the expert group. The first test considered the decisions made by the participants and experts on the scale of 1 through 5. When this test was completed, the experts were again proven to be different from the other pilots (X^2 (5) = 27.58, with 9.49 critical at $p < .05$). The most telling feature of the expert group's performance is that every single expert pilot made exactly the same decision.

Finally, the test of outcomes between the expert pilots and volunteer pilots was calculated. This is the question that your passengers most care about: Will we land safely? Using the X^2 test and the outcome categories of LS, LU, and PA between all participants and the experts produced clear evidence of a difference in the outcomes of the participants and experts ($X^2(3) = 34.91$, with 5.99 critical at $p < .05$). It is significant that every single expert pilot completed the first simulator session with a safe landing. Only 13 of 58 project participants matched that performance by landing safely in their first session.

Understanding the numbers

On the basis of three t-tests (participant groups compared to each other and participant groups each compared to experts using units of time) and six X^2 tests (participant groups compared to each other for decisions made and outcomes and participants compared to experts for decisions made and outcomes), there is strong statistical support for two claims. The first claim is that the two participant groups were selected from the same pilot population and that neither group displayed greater skill with regard to decision time, decisions made, and outcome of the simulated flight. The second claim is that the experts who participated in the project were indeed a different group and displayed much superior decision times, decisions made, and outcomes in the simulated flight. These distinctions are important because they form the foundations from which assumptions made throughout the remainder of the project rely. Any improvement detected within the groups on the second simulator session would therefore be a result of the project and not because one group had a head start on the other.

Once the credibility of the participant groups and experts was established, the focus shifted to the effects of the project on the participants. The second simulator scenario was different from the first, but the essential decision-making elements remained. A decision ranking system was used for the second session, as it was for the first, based on the decisions that the participants were observed to make. The categories used for the second simulator sessions were coded

1 No decision made. Took the extended vector and the holding pattern. Engine failure occurred while in the hold.
2 Late decision made. Took the extended vector but asked for diversions to an airport other than BNA.

3 Early decision made. Did not take the vector but asked for diversions to an airport other than BNA.

4 Late decision made. Took the extended vector but declared an emergency and asked for BNA.

5 Early decision made. Did not take the vector. Declared an emergency and asked for BNA.

Like the first set of decision codes, the second set was also ranked on a continuum with 5 being most favorable and 1 being least favorable. The codes from the first and second sessions were considered to be equivalent. In other words, a category placement of 2 on the decision scale from the first simulator session and a placement of 2 on the scale from the second session would reflect an equivalent level of decision skill. If a participant made a number 3 decision in the first session and later made a 4 decision in the second session, this would be considered an improvement in decision skill because he or she moved along the continuum to a point closer to what is most favorable.

The outcome categories used in the second session remained the same as from the first session: LS for landed safely, LU for landing under unusual circumstances, and PA for a probable airplane accident.

To measure improvements in decision making using the 1 through 5 decision scale, the entire group of participants, both the naturalistic and traditional groups, was first tested by comparing their decisions in the first simulator session with the decision made during the second session. The test employed was a X^2 test using the following logic: The observed decision made by a participant in the first session would be the same as that expected decision in the second session if no treatment effect existed. In other words, there should be no change in decision performance pre- and posttest if nothing special was taking place inside the project. But a change would mean that something within the project did create an improvement. The tests resulted in evidence that a very large improvement had taken place ($X^2(5) = 149.54$, with 9.49 critical at $p < .05$).

A similar test was conducted between the entire participant group pre- and posttest, but this time using the outcomes results LS, LU, and PA. As expected, a large improvement was indicated ($X^2(3) = 35.79$, with 5.99 critical at $p < .05$). This is clear evidence in favor of the assertion that a benefit was received by the participants of the project. The observed outcome of the second simulator session was that 32 of 58 participants landed safely. Recall that only 13 of 58 landed safely during the first session.

The last set of tests attempted to answer the following question: After it was all over, were the pilots from the naturalistic group better than the pilots from the traditional group? When the second session decisions of the traditional group were compared with the second session decisions of the naturalistic group, no significant difference existed between the two participant groups with regard to decision scale ($X^2(5) = 0.442$, with 9.49 critical at $p < .05$). The naturalistic group did make slight improvements, albeit not statistically significant, over the traditional group. The naturalistic group had no participant make a category 1 decision, which is the least favorable and had more participants make decisions in categories 4 and 5, most favorable than the traditional group.

The comparison was also made between the second session outcomes of the traditional group and the second session outcomes of the naturalistic group. Once again, no statistically significant difference was found ($X^2(3) = 1.284$, with 5.99 critical at $p < .05$). Again, although not statistically significant, the naturalistic group had four more participants land safely and two fewer participants end in a probable accident than the traditional group.

What are the numbers telling us?

Taken in combination these tests lend statistical support for these statements:

1 The project produced statistically significant evidence that improvements were made in pilot participant performance with regard to the time it takes to make a decision, the quality of the decision made, and the quality of the outcome produced.
2 The naturalistic group displayed a greater level of improvement with regard to the time taken to make a decision, quality of the decision, and quality of the outcome resulting from the decision, although these improvements were not made to a statistically significant level.

Correlation

Finally, I wanted to see how strong a correlation existed between the units of time that a participant took to make a decision with the outcome from the decision. I used the Pearson *r* coefficient when comparing units of time with the decision made in the first simulator session. The interpretation of coefficient, in its simplest form, divides the variability into two parts. One part is the "explained" variability and the other is the "unexplained" variability. My calculations of the coefficient indicated that 70.2 percent of the variability in decisions made is explained by the time it took to make the decision. The remaining 29.8 percent is unexplained by this relationship. A verbal description of the 70.2 percent value is that a greater number of participants tended to make better decisions when they took less time to make the decision. No causation should be implied here. This calculation does not supply evidence that a shorter decision time causes a better decision; it only reports the relationship. Nevertheless, pilots who were unsure of themselves and did not make timely decisions were much more likely to have an unhappy ending to their flight.

The correlation calculations left a large portion (29.8 percent) of the participant's actions unexplained. It was clear that more was taking place inside the environment of the project than was being explained by mere numbers alone. The quantitative methods employed with the data were important but did not tell the entire story. The shortcomings of the quantitative methods, therefore, pointed the way to an evaluation of the data using qualitative methods to achieve completeness of understanding.

The qualitative approach

To better understand the mountain of data that was collected during the first and second flight scenarios I use many mathematical tests. The statistical analysis of the pilot's performance was effective, but still questions about the pilot's decision behavior remained. Some pilots landed safely even though they had made late or poor decisions. Some pilots concluded with an aircraft accident even though they had made favorable decisions. It became clear that conventional statistics alone would never completely explain what was happening. After all, pilots are humans, and human behavior rarely fits neatly into math problems. So I applied a different kind of analysis to the data. I applied a qualitative approach.

Qualitative research is sometimes referred to as natural method because the researcher must frequent places where events naturally occur, gather data through observations of people engaging in natural behavior, and describe the phenomena occurring in the setting by means of rich, thick descriptions to create an understanding of what is taking place. Situations that lend themselves to this type of analysis take on certain characteristics. Researchers Miles and Huberman in their book Qualitative Data Analysis listed these features of situations that could utilize the natural method:

1 *Prolonged engagement in a field situation,* like many hours observing pilots in flight simulators.

2 *Using the researcher as the instrument.* In my case I was researcher, air traffic controller, and facilitator

3 *Gaining a holistic view of the situation.* After observing the pilots I was able to see past the individual's performance and develop an understanding of the bigger problems that produced the poor performance.

4 *Capturing data from the "inside."* I saw first hand the frustrations, disappointments, and satisfactions of the pilots.

5 *Explaining multiple sources of data.* I was able to use mathematical statistics and also observation, responses from questionnaires of the pilots, and the results of their scenarios to look at the problems from various angles.

6 *Uncovering patterns within the data.* Often pilots would make the same mistakes. These mistakes were not isolated but a symptom of a larger problem. After a while it became easy to see that pilots follow certain habit patterns that either led them into trouble or helped them get out of trouble.

7 *Interpret "soft" data using rich description.* When attempting to understand why pilots did what they did, using words to describe what happened became more helpful than expressing what happened using numbers and mathematical tests.

It seemed evident that the features needed to use a natural method were all present when observing pilots in decision situations. So, in addition to merely starting the timer and recording event times, I also recorded participant actions and quotes. I noted when the participants switched radio frequencies, when checklist items were omitted, when instrument indications were first noticed, what corrective actions were taken when faced with a problem, and even speech patterns and body language. Every pilot's performance in the simulator sessions was recorded. In addition, I and my assistant, Lauren Bandy, kept separate field journals. Over the weeks and months as the project progressed it was clear that participant performance could be clustered, or grouped.

On a normal evening during the project, four sessions were conducted. Using the two Frasca 141 Flight Training Devices, two sessions were conducted simultaneously starting at 5:30 p.m. Then two more sessions would begin at 7:30 p.m. After each evening's sessions and after the participants had left, Lauren and I would have a short conference. It was clear after several weeks that both of us began to see the same recurring themes. The fact that themes began to be observed was not surprising. But we were surprised with the extremes that the themes represented. Routinely, some participants performed extremely well, but just as routinely other participants did so poorly that an unexpected fact emerged: If the participant's performance had been conducted in an airplane in flight rather than in a ground device, the outcome of the flight would have been an aircraft accident with the high probability of being a fatal accident. It was apparent that a combination analysis utilizing both quantitative and qualitative methods was demanded by the project.

Again, using the logic that any analysis method is only a tool and that tools must be properly fitted to the job, we began to rely more heavily on the natural method and less on the statistical method. Michael Patton, in his book *Qualitative Evaluation and Research Methods* (1990), says that this natural method "means that the patterns, themes, and categories of analysis come from the data; they emerge out of the data rather than being imposed on them prior to data collection and analysis." This is precisely what happened in our project. No preconceived categories were developed before the data collection began, but over the course of the project the categories showed themselves. The groupings were of participants who displayed like patterns in their thought process and actions. The first goal was to discover, identify, and code these patterns. The next step was to name the categories in a way that described the category. The groups eventually became the (1) information manager, (2) nonassertive decision maker, (3) snowball effect, and (4) lost in space (Chap. 6).

The data collected during each participant session was one form of primary observation. The field journals kept by Lauren and me were a second primary source, but more data was needed to "triangulate" the data. Only by seeing the same phenomenon from different angles can the phenomenon stand up to scrutiny. A phenomenon observed only once and by only one method cannot be trusted. The phenomenon may be representative of a trend or a pattern, but it also may have been a once-in-a-lifetime fluke. Data that will have a far-reaching impact on a field of study cannot be based on a one-time event that happened to be seen by a researcher. Triangulation of the data ensures that what is observed is validated. For that reason this study did not rely on two types of primary observations alone. The study also incorporated secondary observations in the form of two sets of cued responses. The cued responses were acquired after each of the participants' simulator sessions. The combination of two sets of primary observations and two sets of secondary observations provided the triangulation for the study.

Additional data from flow charts, journals, and questionnaires

After the first session each participant was asked to read a short story that involved a person immersed in a decision-generating situation. The story came with a flow chart of the decisions made by the individual in the story (detailed previously in this appendix). The participants were asked to think of the story as an analogy and to draw out their own flow chart of the decisions they had just made during their simulator session. This produced information of what participants thought was important and how they prioritized the decision situations. The flow chart was an attempt to understand what was going on in the participant's thought process during the session.

The experts who participated in the project were also asked to draw a flow chart of their decision process following their simulator session. These data displayed the thought process of the experts. With this information additional comparisons between an expert's and a nonexpert's decision process was possible.

After the second simulator session each participant was asked to fill out a questionnaire. In the questionnaire the participants were asked to evaluate their own scenario performance. They were asked to determine whether their second performance was better, the same, or worse than their first performance. For those who responded by saying that their performance had improved, the questionnaire asked them to identify a reason for the improvement. This generated data on what each participant perceived was helpful in producing their improvement.

The pilot categories

Information managers

The information managers' flowcharts were all very similar. Each event that took place during the scenario appeared on their flowcharts. In other words, they did not miss any important clues or parts of the scenario. They all appeared to take on each challenge, working it to a logical conclusion. Whenever a solution was not found to a problem, an alternative was decided upon. During the first scenario a red light came on inside the simulator that alerted the pilot to an alternator failure. The pilots in this group all attempted to bring the alternator back on line by turning the alternator switch off and then on. When this solution failed to turn off the light, they moved to an alternative solution, which was to turn off electrical devices that were not absolutely needed to land the airplane. The information manager's flow charts reflected this process. Likewise, when the electronic glide slope also failed, they reasoned that in light of the alternator failure it would be better to attempt an approach even without the electronic glide slope. The approach that was left available to them would not allow the airplane to come as close to the ground as would be the case with the glide slope, but they reasoned it was worth an attempt. The information managers all logically solved the problems or found viable alternatives and *all landed safely*.

The nonassertive decision makers

As mentioned in Chap. 6, the nonassertive decision makers talked during the session much more than the information managers did, but most of the conversation was either to solicit suggestions from the controller or to confirm a decision they were unsure of. The following are example quotes from the nonassertive decision makers:

- We make a missed approach here, right?" (Notice the word *we* when he was the only one in the airplane/simulator.)
- What course of action would you suggest?
- Do we land or go-around? (This comment was made after an hour of flying in the clouds, electrical failures, turbulence, diversion to an unplanned alternate, and a very rough approach to a position under the clouds. After all this he was still unsure whether or not he should land.)

- I want to go to St. Louis, no Ft. Campbell, no Nashville. (This comment was made after the pilot had asked where the nearest VFR/fair weather conditions were located. The scenario called for widespread low clouds back to the leading edge of a cold front that stretched from Tulsa to St. Louis to Chicago. The nearest VFR conditions was behind the front and well out of the airplane's fuel range and battery life without the alternator.)

- What are my instructions? (This question was asked after a missed approach was executed at the destination airport. This is the location where the pilot would tell the controller what he or she intended to do.)

- Well, I have not seen the glide slope for a while, so should I go missed? (The missed approach is the most critical point in the approach procedure and possibly the most dangerous. It is executed when the runway cannot be seen through the clouds even though the airplane is at its lowest safe altitude. Making a missed approach means the airplane is low to the ground, but the pilot cannot see the ground. The missed-approach procedure always starts with a climb away from danger. Any hesitation here could obviously create a ground or obstruction collision hazard.)

- Nashville, I'm missed approach at Smyrna, awaiting further instructions.

- Do you want me to make another approach at Smyrna?

- Maybe I should go to Nashville?

- Approach [control] I can try again or come to Nashville, what do you think?

After each session impressions of what had been seen of the pilots and their performance was written down. After many of these sessions had been concluded, we independently went back through our notes. It was easy to see problems reoccurring among the pilots. Trends were emerging that later helped define and name the pilot categories. These notes were taken from our journals after watching pilots in the nonassertive group:

- More often than not, they will automatically go back after the first failed (instrument) approach to a second approach even when there is no weather improvement and with the knowledge/experience that the first attempt had failed.

Could this be a fall-back response? Because they have no confidence in their own decision, do they fall back to the "over and over" or "one approach after another" routine taught during their flight training?

- Most of these pilots will not tell ATC of the alternator and glide slope problems. They keep them to themselves rather than getting help that could save their lives.

- It seems that when these pilots make the first missed approach and are asked by ATC, "What are your intentions," they have an internal conflict. Could it be that this is the first time they were forced into a decision during a critical time? Every other time did their CFI tell them what to do?

- It would seem illogical that a person would shoot a second approach to an airport that he or she had just flown an approach to and could not get under the clouds while no [weather] improvement was being reported. Maybe they are not illogical; they just do not know anything else to do. They do not know how to make this decision, and they do not know what is possible for them to do. They must be thinking: "So what the hell, maybe the weather will get better."

- When I asked participants why they did not declare an emergency they say (1) fear of FAA, (2) male ego, (3) fear of doing so, and (4) I don't know.

The flow charts that were drawn by this group fell into two categories. Some drew the chart differently than they had actually flown. This group never acknowledged with their chart that they had not been decisive. The other group did acknowledge this fact and some made light of it. One participant drew two charts, one labeled "what I did" and the other labeled "should or would do." One drew a block of the chart with the caption, "continued with unrealistic expectations that weather might improve." One said, "asked approach (control) for help, not much info back."

The snowball effect

Following sessions with pilots in the snowball effect group, I listened to what they had to say. Their comments usually fell into one of two groups. One group wanted to justify their poor performance by blaming the airplane, the controllers, the test, or anything else they can think of:

- This thing should have an autopilot.
- I'm not used to where things are in this simulator.
- I had never flown those approaches before.
- If this would have been for real, I would have done things differently.

The other group seems to be resolved to the fact that the flight was more than they were prepared to handle and to blame themselves:

- I was so rusty, I looked like a beginner in there.
- I got behind the airplane and could not do anything.
- Do you think I would have crashed?
- I needed this, I really needed this.
- Wow, was that a workout.

The captions that these participants wrote into the boxes of their flow charts were also very telling:

- Distracted by problem, got disoriented.
- Tried to get used to lack of feel and instrument discrepancies.
- Ignore frustration that Frasca (simulator) is causing." (This was the participant who ripped out the chart from the book and did not participate in the remainder of the project.)
- Settled down just a bit, still disoriented.
- If [this were] actual [airplane] may have had a mishap.
- Main thought: to fly and keep aircraft under control.
- Distracted from details of the approach, trying to think what to do.
- Did not recognize glide slope failure.
- Get pointed in direction of destination.
- Shot approach at Smyrna (destination airport with clouds too low for a successful approach) because I missed the weather.
- Get ATC with me.
- Have minor electrical problem. (Of course the problem was not minor. Left unattended a complete electrical failure in the clouds was imminent with the possibility of a safe landing unlikely.)
- Approach getting off. Unsure of situation.
- I need a little time.

- Look for problem; don't dwell on it.
- Concerned about loss of glide slope and alternator. Falling behind even further.
- Missed ILS frequency for MQY.
- I was overwhelmed with the simulator and approach.
- I didn't take time to double-check everything.

Two participants wrote in the last box of their flow chart

- Battery dies and pilot
- Set downtown Smyrna on fire.

The lost in space

This group's pilots in general flew the simulator fairly well, but that is all they did. When I initially conceived of the first simulator scenario for the project, it was designed to place the participant in a position that warranted a decision, and the variable to be measured would be the time it took to make the decision. The most unexpected discovery of the entire project was that a large segment of the pilot population never made a decision at all. Worse yet, they failed to make a decision because they were unaware that a decision was desperately needed.

There came a point in the first scenario where all the information required to make a decision had come together. The alternator had failed, and this meant their time was limited and fleeting. This should have been alarming and moved the actions taken into urgency. This group ignored the warning or did not know that it was a warning. Next the clouds were reported at 300 feet above the ground at the destination airport. Three hundred feet is very low and only one type of instrument approach, an ILS, can bring a pilot that close to the ground safely. Then a vital component of the ILS failed: the electronic glide slope. Without this component the airplane on this particular approach could only descent to within 400 feet over the ground. What this group never calculated, or never even realized a calculation was called for, was that the lowest altitude that this approach could reach was still 100 feet higher than the base of the clouds. They were never going to see the runway if that weather report was true. Many pilots from other categories understood that there was a slim chance of landing at the destination but reasoned that they were already at the destination and it was worth a try. Maybe the clouds would actually be higher than reported, and they would see the runway despite the altitude of the cloud bases. These pilots had the missed-approach procedure ready to go and anticipated the need for a missed approach. The lost in space group never made this distinction. They flew the approach down to the lowest safe altitude, expecting to see the runway. They continued their descent through the minimum safe altitude as if they did not know that altitude was important and many descended happily into an airplane crash.

Some never looked at the approach chart, so they could not have known the critical minimum altitude. These pilots were completely lost when it came to position awareness. They "drove" the airplane around when a controller gave them an instruction, but they were oblivious to the big picture. One pilot did level off at the minimum altitude but was unaware that the missed-approach point had been reached. The runway was just under him but obstructed by that last 100 feet of clouds. He flew past the missed-approach point, flying level at the minimum altitude for an additional 6 miles before he decided to climb out. Over the actual terrain there is no obstruction protection past the runway. This pilot would have flown into towers or terrain before he ever got 6 miles away from the airport at the minimum altitude.

Some of these pilots did land safely. But in every case it was not due to pilot judgment or skill that the flight ended safely. When they descended below minimum altitude and eventually reached the base of the clouds some were in a position to see a runway and landed on it. In these cases, the location where they emerged from the clouds was not calculated or being flown with the use of a navigational aid—it was luck.

The lost in space pilots seemed less stressed than others and were quite talkative during the scenario:

- Smyrna Tower I am over the outer marker. (This comment was made as the pilot passed over the middle marker—the markers are 4.3 miles apart and the airplane should have been approximately 1000 feet lower over the middle marker than the outer marker).

- Smyrna tower this is Frasca 141 requesting a low approach. (The reason for flying an approach procedure in the first place is to get under the clouds and land. Why would he ask for only a low approach? He should desperately want to land if this was the flight's intended destination).

- I need to get down.
- Over the outer marker inbound. (This comment was made 5.5 miles outside the outer marker.)
- I almost broke out. (This comment was made to the controller. At this point there was no way the participant could have known how close the airplane was to breaking out from under the clouds.)
- I see the (approach) lights. (Seeing the approach lights means that the runway is in sight and a safe landing can be made; however, when this participant said this, the lights were in fact not visible. The participant lied about what he saw.)
- Am I dead yet? (This comment was made 6.4 miles past the approach runway and 300 feet lower than the minimum altitude. The answer was Yes.)

And one participant said when the scenario had concluded, "This is a different world."

Many of the flow charts drawn by the members of the lost group did not include key elements of the scenario. Quotations from the flow charts included:

- I could not remember exactly what to do.
- I know I missed something.
- I knew I should do something but not sure what.
- Simulator malfunction. (I was never aware of any simulator malfunction.)
- Messed up approach. Need to do something, but what?
- Alternator loss reported, but I did not take action.
- Not sure what to do.
- Noticed DME went out. (I never caused the DME to fail nor was it a part of the scenario. The participant had tuned in an incorrect frequency.)
- Should have powered-off unnecessary equipment.
- Results: poor approach with possible accident.
- Fail to descend fast enough.
- Did not notice alternator.
- Incorrect frequency and misread chart for minimums.
- Primarily worked to keep the plane in a stable attitude.

The illogical decision makers subgroup

The participants who were placed into this subgroup tended to take one of two courses of action. Either they would do the right thing for the wrong reason, or they would do the wrong thing using an illogical reason.

One participant, upon hearing that the clouds at the destination airport were 300 feet, declared that the approach could not be flown. At that time the session had not reached the point where the scenario called for the glide slope to fail. So at the time when this participant decided to go to the alternate, a full ILS approach was still available. That particular ILS approach allows the airplane to be flown to within 200 feet of the ground, which would have been under the clouds. The best choice, given the circumstances at that moment, would have been to take the closer approach. If this participant had taken the best choice, she would have ultimately received a glide slope failure, and that would have changed the choices again. In that situation a diversion to the alternate would become the best choice. So, this participant did do what was considered best but arrived at the decision accidentally through faulty logic.

Another example of a participant making a good choice for the wrong reason came about when one participant declared a missed approach at the first sign that there was a problem. He said, "I was trained to go to the alternate if anything ever failed." This impulsive reaction defies the logical question: Why would the failure be any better at the alternate? This reaction did not take into consideration any of the situation-specific circumstances. In this case the participant did save time by making a missed approach early, but a one-size-fits-all solution would not work in every situation.

The remainder of the participants within this subgroup were characterized by making illogical decisions that placed them in additional jeopardy. Several of the participants would fly one approach procedure and see for themselves that it was not possible to safely land from that approach. Then, with time running out, they elected to fly the same approach again, with the same cloud base weather reported. The time it took to fly the approach a second and sometimes a third time depleted the battery and eliminated all other options that they would have had if they had acted sooner.

As alluded to before, this action may have been due to the fact that they were simply overworked and could not think of anything else to do. The solutions to the problem that were available did require some thought and even creativity. The participants who were mentally taxed by the demands of flying the simulator seemed not to have the time to give the problem any thought, so they reacted with a no-thought action. The product of a no-thought action was an illogical action. This is why the illogical decision makers were considered a subgroup of the snowball effect category.

It was also possible that participants elected to take a no-win course of action because they were simply too timid to work out any other course of action with the controller. Many participants showed reluctance to converse with the controller, and they did not realize that ideas that they might have to solve the problem were valid. They did not seem to realize that they could be and should be in control of the situation. For this reason the illogical decision makers were also considered a subgroup of the nonassertive decision makers category. The requests some of these pilots made became a window into their faulty decision making:

- I'm going around off the approach. (This comment was made at a point when the approach had been going well and landing was a real possibility.)
- Nashville (approach) I am requesting an ASR [Airport Surveillance Radar] approach. (The participant had just made a missed approach on a procedure that allowed the airplane to descend to within 400 feet of the ground. The ASR approach is a procedure that allowed the airplane descent to only 600 feet.)
- I was close to the airport, so the alternator light was not urgent.
- The weather is below localizer minimums. If I miss this approach, I will return for another at Smyrna.

Our journals again turned up surprising trends:

- The decision-making process seems to be independent of how current an instrument pilot is. Some of our subjects made good decisions yet flew the Frasca 141 poorly; others flew well but made poor decisions.
- There seems to be a universal fear of seeking help or declaring an emergency. They risk their lives rather than writing a letter or making a phone call. (When a pilot declares an emergency, the FAA will sometimes ask the pilot

to explain what happened in a letter or telephone interview. See Chap. 8).

The flow charts that were drawn by this group also yielded evidence of faulty logic:

- Asked to go back around (to the same instrument approach) because not sure what to do.
- Request/receive ASR.
- Decide to try another ILS; if glide slope inoperative, request localizer approach clearance.
- Declare missed and try again localizer approach only.
- Asked for revector to ILS 32 (the destination airport).
- Decided to take a look and pray for a hole.

Good decision makers/poor fliers subgroup

This group was the smallest of the groups identified, indicating that the physical control of the simulator/airplane is much less of a problem overall than the mental ability to reason out problems and anticipate.

Two participants of the 65 pilots to fly the first scenario were unable to control the simulator well enough to complete the scenario. These two lost control and crashed before any challenges from the scenario were produced. The majority of the poor fliers group, however, were able to complete the scenario, but their lack of airplane control skills prevented a successful outcome. The most common occurrence within this subgroup was a pilot-induced missed approach that forced the airplane to remain in the air longer than the electrical system or engine lasted. In other words, if they had flown the approach well, they would have landed safely. Their decisions brought them to the correct place, but they could not execute the plan. Their mental work succeeded, but their physical work failed.

Another characteristic that emerged from this group was that poor flying actually influenced future decisions that were made. When a procedure was flown poorly, the pilot did not get the same information that he or she would have received had the procedure been flown correctly. An example is a pilot who did not descend all the way down to the altitude allowed on an approach. Flying all the way down would have shown that the clouds were still too low for the approach and another attempt would be a waste of time. But having never gotten that low, this information was not available to them, so often they would attempt the approach a second time and in doing so reduced their options. The second attempt was a poor choice, but it was made without the information that might have shown the pilot that it was a poor choice. In this sense a correctly flown approach that does not end in a safe landing is still not a total loss, because the pilot, although not gaining a landing, did gain information that could be used in the next decision. So a poorly flown procedure robbed the pilot of vital information necessary to achieve an ultimate solution. A poorly flown procedure became a double-edged sword. The struggle deprived the pilot of both aircraft control and information, both of which were needed for a favorable outcome.

The good decision/poor fliers subgroup made these representative comments during the session:

- Full-scale needle deflection, I'm going around. (The full-scale needle deflection indicated that the airplane was off course to such a large degree that the instrumentation could no longer identify position.)

- Now I have lost the glide slope again. (The participant was flying an approach that had an operating glide slope, but flew the approach too high and overshot the proper descent path. The error became so large that the airplane's instrumentation could no longer identify position.)

- There it goes again. (The participant was commenting on the fact that the needle that defined the approach course was swaying back and forth rather than holding steady. This was happening because the pilot was inadvertently making turns across the course.)

The participants in this group were completely aware that their poor aircraft control and inability to fly the procedures with precision had cost them a safe outcome. The comments made on their flow chart included

- Second try too high; go around.
- Essential to make best possible approach.
- High; go around.
- Decided I was too high on first approach. Second time get down sooner.
- Misread localizer (approach). Thought it was inoperative, so I asked for a vector back for a second localizer.

- Pilot does poor job of navigating ILS and flies through glide slope. Time runs out. Battery dies and plane crashes on airport.

Declaring an emergency affected safety

In Chap. 7 the fact that pilots are reluctant to declare emergencies was described. Many of the volunteer pilots came to the sessions for the first time thinking that the whole project was some sort of undercover FAA trap. This attitude was so universal among the pilots that I started explaining that the project was only research immediately after I introduced myself. They became much more comfortable when they started to believe that I was a pilot just like them, and my goals were to get information for better training not get information on them. I sent out 1189 invitations to join the project. Of those I ultimately received 139 response cards in return. I will always suspect that of the 1050 pilots who did not respond, some did not send the card back because they thought it was some kind of FAA sting operation. So in a way, the FAA's poor reputation cost me and you valuable information that we might otherwise have had.

Chapter 7 also describes the FAA process and forms used with pilot who declare emergencies. During the workshops I showed these forms to both the traditional and naturalistic groups. The message during the seminars was that the FAA's prosecution of a pilot who declares an emergency is rare, and even if it were common, it would be better to be in trouble after landing than to be involved in potentially fatal aircraft accident. During the second simulator scenario, there came a point where the declaration of an emergency was an option, and it could be argued that it was the best option. In the second scenario the pilot was flying along with a rough-running engine on the way to Nashville International Airport. The cloud ceiling was reported well above the approach minimums. Then the pilots were told that another airplane landing at Nashville had just made a gear-up landing and emergency crews were en route to the scene of that accident. Because the emergency crews were on the scene of an accident rather than at their ready positions, the airport was closed. The entire airport was closed even though the accident did not block the runway that the pilots were being vectored to. The pilots in the scenario were given a holding pattern and asked to wait until the airport was reopened. The pilot's engine continued to run rough and worsen. The oil pressure gauge was low and on the red line by now, and the oil temperature gauge was high and on the red line. Each time a pilot asked about the airport availability, the controller told them that the airport was still closed and to remain in the holding pattern.

Thirty-five of fifty-eight (60 percent) of the participants did use the declaration of an emergency to help solve the in-flight problem that they faced. Of these 35 participants who declared an emergency when one was warranted, 33 eventually landed without a probable accident. The pilots in the scenario had not violated any rules. They had done nothing wrong. But they did face a set of circumstances that required immediate, common sense action. One pilot said, "I know you have one accident down there, but you are going to have another one out here if I don't get down. This is an emergency and I'm coming in." The pilots were told that the airport was closed, but they had the power to open it again with a single word: emergency.

The project pilot's mismatches

Many of the participants, when faced with a situation in which a decision was called for, made no decision at all. Other participants made some decision, but there was a greater likelihood of a favorable flight outcome when the decision that was made came from a correct pattern match. As the volunteer pilots came to fly the simulators through the project, I began to see more and more of the matches and mismatches occurring. Only a few minutes into the second simulator session the participants were presented with a diminishing engine oil pressure reading. The oil pressure indication eventually reached the lower red line of the indicator (Fig. A-20), which meant that the engine no longer had lubrication from oil. This was followed in sequence by a rising engine temperature reading. Eventually the engine oil temperature and cylinder head temperature indicators read at the red line (Fig. A-20), which meant that the engine had reached a dangerous level of overheat. The logical conclusion from these two indications is that the engine's lubricant was leaking out, and an engine failure was imminent. These engine instrument indications were followed by intermittent engine roughness. Of the participants who flew the second session, 27 noticed these engine instrument readings and made a report of them to the air traffic controller. Of these 27, only 6 (22 percent) completed the session with a probable accident outcome. In other word, the pilots who were aware of the true situation that they were immersed in made better decisions and flew safer flights. Figure A-22 breaks down the rough-running engine scenario into associated groups. If the pilot had used all the available information, the conclusion of oil pressure loss followed by engine failure would have been apparent.

Of the participants who flew the second session, 12 pulled the carburetor heat valve (Fig. A-23) on when faced with the engine roughness. Carburetor heat is applied when ice is suspected in the engine. Carburetor ice can produce engine roughness and eventual engine failure, but indications of dangerously low oil pressure together with dangerously high engine temperatures are not indicators of carburetor ice. In other words, given the available information, carburetor ice was not a logical reason for the engine roughness. Pulling the carburetor heat valve on displayed a troubleshooting error, which led to a pattern mismatch. Of the 12 participants who made this error, seven (58 percent) concluded the session with a probable accident outcome. The pilots who did not know the true situation tried to solve their problems with remedies that did not match the situation. Pulling carburetor heat to solve a problem of engine oil does not match. More than half the pilots who made this mismatch crashed.

The proper pattern match in this situation created a safer situation for the pilot. If pilots saw the engine instrument readings and understood their implication, they would also know that there was nothing that a pilot could do to remedy the situation while in flight. It is impossible, in the type airplane represented by the simulator, to add engine oil to the engine while in flight. It would also be unlikely for a pilot to have additional engine oil within reach while piloting the airplane. Therefore, the logical conclusion to this situation is that the engine will eventually stop, and the only course of action the pilot has remaining is to get on the ground before engine failure occurs. Properly diagnosed, the pilot should have realized that it was a race against time. Could the pilots conceive of and execute a plan that would get them on the ground as fast as possible? The majority of participants (21 of 27) who understood the gravity of the problem and matched their actions properly concluded the session with a safe landing.

A pattern mismatch in this situation, however, created a much more dangerous condition for the pilot. In reality the participants had no time to waste because of the lack of engine oil. But those participants who incorrectly thought that the problem was not oil loss but rather

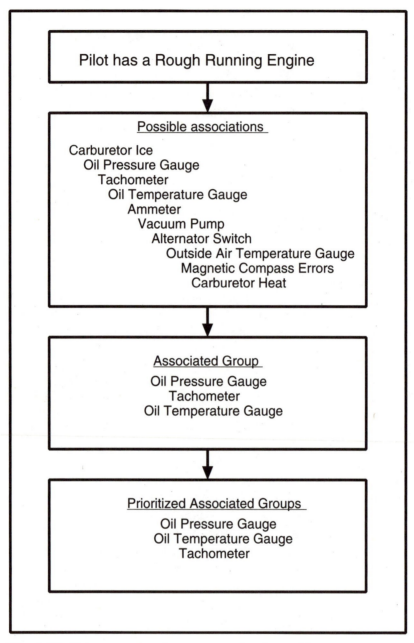

Fig. A-22 *Using associated groups to solve the problem*

Fig. A-23 *The "carb heat" control knob. Pulling the knob out allows heated air into the carburetor in an attempt to melt carburetor ice.*

carburetor ice took their actions in the wrong direction. In a circumstance where carburetor ice was indeed present, the application of carburetor heat would begin to solve the problem, but this process is slow. The ice would melt from the engine over time. Over the time the ice is melting the situation is improving. During the second simulator session the oil loss produced a condition where the problem was deteriorating with every second. But those participants who misdiagnosed the problem as carburetor ice and applied carburetor heat as the remedy thought the problem was improving every second. Their actions then seemed to suggest that they thought they had bought themselves some time. After pulling open the carburetor heat control, and believing that they had solved the problem, they were more likely to choose a course of action that would keep them in the air longer, such as selecting a more distant alternate airport.

Fifty-eight percent of those who pulled the carburetor heat control eventually ended the session in a probable accident. This suggested that when the participants made a pattern mismatch, they wasted time that otherwise could have been used to land as soon as possible and produce a favorable outcome. The pattern mismatch in this circumstance set the pilots' decision process on a diverging course from reality and proper problem solution. The mismatch set the pilot off on a course that made the assumption that things were getting better when in reality a proper pattern match would have created an understanding that things were getting worse. The pattern mismatch created a complete loss of situation awareness. Past that point the situation always became more dangerous for the pilot, and it appeared difficult for them to regain awareness once a falsehood was believed. Decisions made after the loss of awareness occurred only compounded the problem. Pilots often found themselves flying away from an airport as the engine stopped.

The reason the pattern mismatch occurred was the participants' inability to use or understand all the facts that were presented. Many participants later told me that they never noticed the engine instruments. Others reported the engine instrument indications but were not able to interpret their message correctly. The deeper reason that participants were failing to use or understand information that presented itself tended to be preoccupation with other tasks. Of these other tasks the demands of flying the airplane appeared to be the chief cause. Many participants could fly the simulator very well. They were able to hold a heading, hold an altitude, and follow a course. But when the situation progressed to a point where decisions were demanded, the flight control suffered. It appeared that many participants could fly the airplane or think about the problems, but not both. This of course has serious implications for single-pilot operations.

The final cued response

The participants were asked to complete a questionnaire after the second session. The first question was "Do you think your performance in the flight simulator the second time was an improvement over the first time?" The participants were then asked to circle either "Yes," "No," or "About the same." Twenty-seven of thirty-one participants from the naturalistic group answered that they had improved. Four indicated that their performance was about the same. The 27 who thought they had improved were then asked what they attributed their improvement to. Of these, 24 said they had improved by selecting the clearer-understanding-of-decision-making response.

Twenty-four of twenty-eight participants in the traditional group responded that they had improved in the second session. The reasons that they gave for the improvement were split among clearer understanding of decision making, more familiar with the simulator, and staying ahead of the airplane.

The following is a list of responses to the question What did you learn from this project that you can use to make your next flight safer?

- Stay calm, prioritize the events in your flight, think ahead, use all your resources.
- Use all available resources. When I'm overworked with flying the aircraft use controllers, center, etc.
- Put ATC to work for me.
- I learned the value of staying ahead of the airplane.
- Taught me a different attitude to use.
- IFR proficiency is a never-ending goal.
- The importance of thinking the situation through. Consider every possible alternative.
- Consider your options; then making decisions should become a much faster process for me.
- Be more assertive—get going—ask for help at the early stages.
- Better thinking.
- Handling stress management. Making clear-cut decisions.
- An appreciation of situation awareness (how critical it is to the safety of flight). The simulator sessions really made this real world applicable."
- I'm sure this took me a step above being only "approach certified."
- Don't be afraid of ATC.
- *It can* happen to me.

- Consider all options available.
- Think ahead and be informed of the situation during the flight by both the controllers and the airplane.
- Try to leave yourself an out.
- It encourages me to become more current.
- Don't get tunnel vision.
- I am always in control of where the airplane flies and lands (not the controller).
- I learned that I must always be learning.
- Although I knew enough to pass the IFR checkride, I desperately lack the comprehensive skill and knowledge required to keep me alive in an emergency. I need to really think through some situations to prepare myself better for the out-of-the-ordinary situations that I will surely face. Craig's [the author's] point that rote training for the checkride is no preparation for the real world is well taken.
- Rational decision making will keep you cool.
- Be more proactive; use all resources available to you, including ground controllers.
- I think this will change how I approach instrument flying.
- Trust my own judgment. Remember who is in control.
- The [naturalistic] lecture made a huge difference in my attitude. That needs to be part of every biannual flight review. You need to get FAA to add this to recurrent training.
- It improved my understanding of the need to put decision making on a priority basis for sequencing.
- To be more expectant of problems and take command and be responsible for my own flight.
- In the past few months my IFR flying was becoming more of a physical activity, sort of a "rote" type activity. This course helped me to move back to a more "vertical" type flying. The ASAP card helped. I have it mounted on my instrument panel.
- Very worthwhile. Probably saved some folks from themselves.
- Make decisions and trust your instincts.
- Take control. Stay ahead of the airplane—this is big.

- This could easily save my life (and those with me) someday.
- Be assertive when necessary.
- Analyzing options is a skill that must be maintained.
- No such thing as too much preflight planning.
- Read what the instruments really say, not just what you want them to say.

And another pilot participant concluded, "This project may save some lives. Maybe mine."

Question 4 asked the participants, "Should a 'real-world' emphasis be included in IFR recurrency training?" Most, 94.7 percent (54 of 57 responding), circled "Yes." Several circled "Yes" with multiple circles, which I took to mean an emphatic Yes. Two of 57 participants circled "No," and one circled "Not Sure." Question 5 asked the pilots "Should a 'real-world' emphasis be included in IFR initial training?" To this question 56 of 57 pilots responding, or 98.2 percent, circled "Yes," with one circling "Not Sure."

The last question was a broad opportunity for the pilots to give feedback about the study by asking, "What is your overall impression of the project?" Some of the pilot answers were

- For me it was very helpful. Beneficial to any pilot to be put into "real-world" situations.
- It was fun; it gave me a little more confidence in myself.
- Taught me a different attitude.
- Excellent. Learned a great deal. It was eye opening.
- Definitely made me think. Hopefully I will carry increased awareness with me.
- I had a great time, and I know I improved my flying.
- I needed it bad. I got more than I expected. It helped very, very much.
- Helped me with my decision-making process and to plan my flight thinking about all possible encounters.
- Overall this project will help aviation safety. Instructors should use "real-life" situations in training and recurrent training.
- This project made me aware of my level of competency and not only aircraft control but also of adverse situations and what I can do to survive them.

- It was *extremely* beneficial for me. I enjoyed the challenging situations—they showed me some areas I need to work on. I'd like more real-life situations in my training.
- Needs to be expanded to a full-time course. This takes training in a different direction.
- It was very enlightening. This pointed out that analyzing options is a skill that must be maintained.
- Excellent. I wish something of this nature had been a part of earlier training.
- I appreciate this "Thinking" training. I'm better prepared as a pilot because of this.
- Gives a new perspective.
- It was a great learning process for me, and I think it will change how I approach flying.
- Very good. Please continue this effort.
- I think this was a very beneficial project and feel something of this nature should be required or used more often.
- I believe that I will be a safer pilot because of this.
- A+; I wish I had a regular session with similar "real-world" scenarios at least every 6 months. It should be required for recurrency.
- Everyone should have this opportunity.
- I learned a lot and had fun. That's why I fly!

As you can see the pilots had changed from skeptical to receptive about the project. Many of these pilots stay in touch with me and are always asking when the next project will start. I believe that pilots and instructors should come away from reading this with the impression that given a nonthreatening environment, pilots will be eager to learn and improve their skills. This fact has given me great encouragement and is the reason for writing this book.

Suggested Readings

Accardi, T. C. (1991). *Aeronautical Decision Making*. Washington, D.C.: Federal Aviation Administration.

Amalberti, R., & Deblon, F. (1992). Cognitive modeling of flighter pilot aircraft process control: A step toward an intelligent on-board assistance system. *International Journal of Man-Machine Systems*. 36, 639–671.

Amel, E. L. (1995). "Understanding Pilot Decision Making: Theory and Application." UMI Dissertation Abstracts.

Beach, L. R., & Potter, R. E. (1992). The prechoice screening of options. *Acta Psychologica*. 81, 115–126.

Beaudette, D. C. (1990). *Line Operational Simulations: Line-Oriented Flight Training, Special Purpose Operational Training, Line Operational Evaluation*. Washington, D.C.: Federal Aviation Administration.

Berlin, J. I., Gruber, E. V., Holmes, C. W., Jensen, P. K., Lau, J. R., Mills, J. W., and O'Kane, J. M. (1982). *Pilot Judgment Training and Evaluation* (DOT/FAA/CT-82/56). Daytona Beach, Fla.: Embry Riddle Aeronautical University. (NTIS No. AD-A117 508.)

Besco, R. O., (1995). Improving takeoff abort decisions/performance. *Proceedings of the Eighth International Symposium on Aviation Psychology* (pp. 1336–1340). Columbus, Ohio: The Ohio State University.

Bogdan, R. C,. & Biklen, S. K. (1992). *Qualitative Research for Education—An Introduction to Theory and Methods*. Needham Heights, Mass. Allyn and Bacon.

Bovier, C. (1997). Situational awareness, key components of safe flight. *Flying Careers*. JA. 97. 23–27.

Brehmer, B. (1990). Strategies in real-time, dynamic decision making. In R. M. Hogarth (ed.), *Insights in Decision Making* (pp. 262–279). Chicago, Ill: The University of Chicago Press.

Bruggnik, G. M. (1997) A changing accident pattern. *Airline Pilot, The Journal of the Airline Pilots Association*, 66 (5), 10–13.

Buch, G., & Diehl, A. (1984). An investigation of the effectiveness of pilot judgment training. *Human Factors*. 26 (5), 557–564.

Cannon-Bowers, J. A., Salas, E., & Pruitt, J. S. (1996). Establishing the boundaries of a paradigm for decision-making research. *Human Factors*. 38 (2), 193–205.

Carroll, L. A. (1992). Desperately seeking SA. *TAC Attack* (TAC SP 127-1) 32, 5–6.

Chappell, S. L. (1995). Managing situation awareness on the flight deck or the next best thing to a crystal ball. *NASA Directline.* 95. Washington, D.C.: National Air and Space Administration.

Crandall, B., & Calderwell, R. (1989). *Clinical assessment skills of experienced neonatal intensive care nurses.* Yellow Springs, Ohio: Klein Associates Inc. (Final Report prepared for the National Center for Nursing, NIH under Contract no. 1 R43 NR01911 01.)

Dawes, R. M. (1971). A case study of graduate admissions: Application of principles of human decision making. *American Psychologist.* 34, 571–582.

Dunn, O. J., & Clark, V. A. (1974). *Applied Statistics: Analysis of Variance and Regression.* New York: John Wiley & Sons Harper Publishers.

Dyson, G., & Jimenez, M. (1994). Human Factors Training. Federal Express Flight Training.

Edwards, W. (1962). Dynamic decision theory and probabilistic information processing. *Human Factors,* 4, 59–73.

Endsley, M. R. (1995). Toward a theory of situational awareness in dynamic systems. *Human Factors,* 37 (1), 32–64.

Endsley, M. R., & Smith, R. P. (1996). Attention distribution and decision making in tactical air combat. *Human Factors.* 38 (2), 232–249.

Federal Aviation Administration (1999). *Aeronautical Information Manual (AIM).*

Flathers, G. W., Giffin, W. C., & Rockwell, T. H. (1984). A study of decision-making behavior of aircraft pilots deviating from a planned flight. *Aviation, Space, and Environmental Medicine.* 53 (10), 958–963.

Flight Training Devices Qualification (1992). FAA Advisory Circular 120-45A.

Frasca, R. (1997). PCATDs Counterpoint—Industry needs to be cautious in its approach to computer training devices. *Airport Business, MA 97,* 15–17.

Frasca, R. (1997). The Internet: Frasca Internationals position on PCATDs. http://www.frasc.com/pcatd2.html.

Giffin, W. C., and Rockwell, T. H. (1984). Computer-aided testing of pilot response to critical in-flight events. *Human Factors,* 26 (5), 573–581.

Hart, S. G., & Bortolussi, M. R. (1984). Pilot errors as a source of workload. *Human Factors.* 26 (5), 545–556.

Helmreich, R. L., Wilhelm, J. A., & Gregorich, S. E. (1988). *Notes on the concept of LOFT: An agenda for research.* NASA/University of Texas Technical Report 88-4.

Jensen, R. S. (1982). Pilot judgement: Training and evaluation. *Human Factors,* 24 (1), 61–73.

Jensen, R. S., & Benel, R. A. (1977). *Judgement evaluation and instruction in civil pilot training.* (Report No. FAA RD-78-24) Champaign, Ill.: University of Illinois. (NTIS No. AD-A057 440.)

Kaempf, G. L., & Orasanu, J. (1997). Current and future applications of naturalistic decision making in aviation. In Zsambok, C., & Klein, G. A. (eds.), *Naturalistic Decision Making.* Hillsdale, N.J.: Lawrence Erlbaum Associates.

Klein, G. A. (1995). A recognition-primed decision (RPD) model of rapid decision making. In Klein, G. A., Orasanu, J., Calderwood, R., & Zsambok, C. E. (eds.). *Decision Making in Action: Models and Methods.* Norwood, N.J.: Ablex Publishing Corporation.

Klein, G. A., & McNeese, M. D. (1993) The Internet: Naturalistic decision making: Implications for Design. http://www.dtic.dla.mil/iac/cseriac/qul94ndm.html.

Klein, G. A., Orasanu, J., Calderwood, R., & Zsambok, C. E. (1995). *Decision Making in Action: Models and Methods.* Norwood, N.J.: Ablex Publishing Corporation.

Larkin, J., McDermott, J., Simon, H. A., & Simon, D. P. (1980). Expert and novice performance in solving physics problems. *Science.* 20 (208), 1335–1342.

Law, J. R., Helmreich, R. L., & Wilhelm, J. A. (1994). Key ingredients of high-quality LOFT: A recipe for effective pilot training. *The CRM Advocate.* 94-3, 2–9.

Lester, L. F., & Bombaci, D. H. (1984). The relationship between personality and irrational judgement in civil pilots. *Human Factors.* 26 (5), 565–572.

Lincoln, Y. S., & Guba, E. G. (1985). *Naturalistic Inquiry.* Newberry Park, Calif.: Sage Publications, Inc.

MacDougall, R. (1922). *The General Problems of Psychology.* New York: New York University Press. (Cited in Gillis, J., & Schneider, C. (1966).

McMillan, J. H., & Schumacher, S. (1993). *Research in Education: A Conceptual Introduction,* 3d ed. New York: Harper Collins College Publishers.

Mancuso, V. (1997). Managing Crew Situational Awareness. *Ninth International Symposium on Aviation Psychology, Columbus, Ohio:* The Ohio State University.

Marshall, S. P. (1995). *Schemas in Problem Solving.* Cambridge, England: Cambridge University Press.

Miles, M. B., & Woods, D. D. (1994). *Qualitative Data Analysis: An Expanded Sourcebook.* Newbury Park: Sage Publications.

Miller, T. E., & Woods, D. D. (1994). The Internet: Key issues for naturalistic decision-making researchers in system design. http://esel.eng.ohiostate.edu: 8080/"keyissNatDecBeh.html.

Murray, S. R. (1997). Deliberate decision making by aircraft pilots: A simple reminder to avoid decision making under panic. *The International Journal of Aviation Psychology.* 7 (1), 83–100.

Nall Report, Accident Trends and Factors for 1995. Frederick, Md.: AOPA Air Safety Foundation.

National Business Aircraft Association Management Guide (1997).

Ninth International Symposium on Aviation Psychology program guide (1997). The Ohio State University, Columbus, Ohio.

Orasanu, J. (1998). NASA Ames Research Center. Telephone interview.

Orasanu, J., & Connolly, T. (1995). The reinvention of decision making. In Klein, G. A., Orasanu, J., Calderwood, R., & Zsambok, C. E. (eds.), *Decision Making in Action: Models and Methods.* Norwood, N.J.: Ablex Publishing Corporation.

Orasanu, J., & Fischer, U. (1997). *Finding decisions in natural environments: The view from the cockpit. In Zsambok, C., & Klein, G. A. (eds.), Naturalistic Decision Making.* Hillsdale, N.J.: Lawrence Erlbaum Associates.

Patton, M. Q. (1990). *Qualitative Evaluation and Research Methods.* Newberry Park, Calif.: Sage Publications, Inc.

Personal Minimums Checklist (1996). Aviation Safety Program. Washington, D.C.: Federal Aviation Administration.

Personnel Cockpit/Crew Resource Management Program (1994). United States Air Force. Instruction number 36-2243.

Quality Crew Resource Management: A Paper by the Human Factors Group of the Royal Aeronautical Group (1996). London, England: United Kingdom, Civil Aeronautics Authority.

Schwartz, D., (1993). Training for Situational Awareness. *Airline Pilot.* May 1993, 20–23.

Sexton, B., Marsch, D. P., Helmreich, R. L., Betzendoerfer, D., Kocher, T., & Scheidegger, D. (1996). *Jumpseating in the Operating Room.* Basel, Switzerland: University of Basel.

Simmel, E. C., & Shelton, R. (1987). The assessment of nonroutine situations by pilots: A two-part process. *Aviation, Space, and Environmental Medicine.* 58, 1119–1121.

Smith, D. E., & Marshall, S. (1997). Applying Hybrid Models of Cognition in Decision Aids. In Zsambok, C., & Klein, G. A. (eds.), *Naturalistic Decision Making.* Hillsdale, N.J.: Lawrence Erlbaum Associates.

Starter, N. B., & Woods, D. D., (1991). Situational awareness: A critical by ill defined phenomenon. *International Journal of Aviation Psychology.* 1, 45–57.

Sternberg, R. J. (1985). Teaching critical thinking, Part 1: Possible Solutions. *Phi Delta Kappa,* 67 (4), 194–198.

Sternberg, R. J. (1985). Teaching critical thinking, Part 2: Are we making critical mistakes? *Phi Delta Kappa,* 67 (3), 277–280.

Stokes, A. F., Barnett, B., & Wickens, C. D. (1987). Modeling stress and bias in pilot decision-making. In P. Rothwell (ed.), *Proceedings of the Human Factors Association of Canada,* annual conference (pp. 45–48).

Strauss, A. L., & Corbin, J. (1990). *Basics of Qualitative Research— Grounded Theory Procedures and Techniques.* Newberry Park, Calif.: Sage Publications, Inc.

Sumwalt, R. L., & Watson, A. W. (1995). "Oh no! We've got a problem"— What ASRS incident data tell about flight crew performance during aircraft malfunctions. *Proceedings of the Eighth International Symposium on Aviation Psychology.* Columbus, Ohio: The Ohio State University.

Wilhelm, J. A., & Helmreich, R. L. (1996). A short form for evaluation of crew human factors skills in line flight settings. *The University of Texas Aerospace Crew Research Project: Technical Report 96-6.*

Witte, R. S. (1993). *Statistics,* 4th ed. Orlando, Fla.: Holt, Rinehart and Winston, Inc.

Woods, D. D. (1990). The Internet: Challenges for models of "naturalistic" decision behavior. http://esel.eng.ohio-state.edu:8080/"csevchalmod-NatDecBeh.html.

Zsambok, C., & Klein, G. A. (1997) *Naturalistic Decision Making.* Hillsdale, N.J.: Lawrence Erlbaum Associates.

Index

About the Author

Paul A. Craig is the author of six books published by McGraw-Hill. He is a flight instructor and has been an active pilot for 25 years. His former flight students are today captains at five major airlines, fourteen regional air carriers, over twenty corporations, and every branch of the military. Craig is a former Flight Instructor of the Year in both North Carolina and Tennessee. He currently is the Chief Pilot and Professor of Aerospace at Middle Tennessee State University. Craig has a Doctor of Education degree in curriculum and pilot decision making. He is an airline transport pilot, an instrument and multiengine flight instructor, and recently became a seaplane pilot. Craig regularly speaks around the country at pilot proficiency programs and flight instructor refresher clinics. He is married to Dr. Dorothy Valcarcel Craig and they have two children. The Craigs live in Franklin, Tennessee.